The State and the Farmer

by Liberty Hyde Bailey

with an introduction by Roger Chambers

This work contains material that was originally published in 1911.

This publication was created and published for the public benefit, utilizing public funding and is within the Public Domain.

This edition is reprinted for educational purposes and in accordance with all applicable Federal Laws.

Introduction Copyright 2017 by Roger Chambers

Self Reliance Books

Get more historic titles on animal and stock breeding, gardening and old fashioned skills by visiting us at:

http://selfreliancebooks.blogspot.com/

Introduction

I am pleased to present yet another title on Gardening.

The work is in the Public Domain and is re-printed here in accordance with Federal Laws.

As with all reprinted books of this age that are intended to perfectly reproduce the original edition, considerable pains and effort had to be undertaken to correct fading and sometimes outright damage to existing proofs of this title. At times, this task is quite monumental, requiring an almost total "rebuilding" of some pages from digital proofs of multiple copies. Despite this, imperfections still sometimes exist in the final proof and may detract from the visual appearance of the text.

I hope you enjoy reading this book as much as I enjoyed making it available to readers again.

Roger Chambers

To

ISAAC PHILLIPS ROBERTS

—FARMER, TEACHER, PHILOSOPHER
AND FRIEND—

I offer this book

EXPLANATION

THE immediate basis of this book is a paper on "The State and the Farmer," originating as a presidential address before the Association of American Agricultural Colleges and Experiment Stations at Lansing, Michigan, May 28, 1907. Five hundred copies of this address were printed by order of the Association. The demand continuing, I decided to expand the principles there expressed, and to apply them to a somewhat wider range; in doing this I have, of course, much increased the size of the original paper, but the general motive is not different. That paper, in turn, was the result of previous papers and addresses, some parts of which were incorporated in it; therefore, some of the arguments and points of view in this book are likely not to appear to be new to those readers who have followed this class of discussions. It is needless to say that the book is in no sense a treatise, but only a budget of opinions; and I

Explanation

am well aware how likely it is for opinions to be colored by the particular set of conditions under which one works.

L. H. BAILEY

ITHACA, N. Y.
MAY, 1908.

OUTLINE

	PAGES
The Argument	1–4
Governmental interference	2
I. THE AGRICULTURAL SHIFT	5–54
THE GEOGRAPHICAL SHIFT IN RURAL OCCUPATIONS	5–11
The apparent reaction	8
THE SHIFT IN AGRICULTURAL METHOD	11–15
THE SHIFT IN RURAL INSTITUTIONS	15–21
THE "ABANDONED FARM" AS AN ILLUSTRATION OF THE AGRICULTURAL SHIFT	22–54
The illusion of old buildings	25
Old fields	29
The significance of the general situation	30
The situation with individual farms	35
Lack of adaptation	39
Point of view as to remedies	42
The outlook for the hills and remote lands	51

Outline

	PAGES
II. SOCIETY AND THE FARMER	55–177
THE PROBLEM	56–73
Saving our resources	56
The social questions	61
The countryman	66
Rural needs	69
THE NATURE OF THE SOCIAL REMEDIES	73–176
The importance of the personal and local initiative	74
Agencies of local communication	75
Reconstructive movements	78
The kinds of help	80
1. THE DISCOVERY AND COLLATING OF LOCAL FACT	81–86
Agricultural surveys	83
The model farm idea	84
2. DEVELOPING PARTICULAR PERSONS FOR COMMUNITY WORK	86–89
3. THE GOVERNMENTAL FUNCTION IN AGRICULTURE	89–111
State departments of agriculture	90

Outline

	PAGES
A new statesmanship	92
Attitude of state governments	93
The state government part and the federal part	95
States rights	102
The coördinating agencies	104
Coördination of educational matters	106
Coördination of agricultural matters	109
4. THE REDIRECTING OF RURAL INSTITUTIONS	111–135
The necessity for working together	114
The economic organizations	116
What agricultural societies can do	118
Possible extent of associative work	122
Rural government	125
Application of public money	126
Banks	128
Fairs	130
The rural church	132
5. THE DEVELOPING OF APPLICABLE EDUCATION	135–172
Necessity of a new point of view	136

Outline

	PAGES
Importance of the rural school	140
(1) *The subject of federal aid* . . .	143
(2) *The consolidating of schools* . . .	147
(3) *Special agricultural schools* . .	150
The redirecting of the rural school . .	157
The agricultural colleges and experiment stations	164
The extension work of the colleges . . .	169
6. APPEAL TO PERSONAL LEADERSHIP	172–176

THE STATE AND THE FARMER

THE person who works his own land for a living is usually a strong individualist. He looks to the earth, rather than to persons, for his livelihood. He does not cater. If, in any country, he patronizes, it is because of his social position, not because he is a farmer.

This individualism conduces to isolation of ideas. The farmer's work is founded on personal experience; and when he is not able to analyze his experience or to understand it, he falls into the experience-routine of the season, and his ideas become crystallized.

The first or original real occupation was the management of land. It is the basic occupation. Out of it most other occupations and trades have developed. The constructive and imaginative spirits took to these newer trades, and the conservative forces tended to remain on the land.

As the demands of civilization have developed, and particularly as world-competition has arisen, the isolation-ideals of the land-worker have been more and more inadequate to meet the conditions. A new type of mind has been forced on him. As community-ideas have evolved, fellow land-workers have assumed new relations to each other. As the community-sense has grown into nationalism, and as loyalty to the person of a local leader or ruler has developed into patriotism, the organization of society—or the government—has felt the necessity of interfering with the land-worker, as with other workers, for the benefit of society at large.

Governmental interference.

With the enlargement of the necessities of mankind, and the organization of society, therefore, the land-worker has been pressed by two opposite and somewhat opposing forces,—the necessity of improving his own practice, and the necessity of being compelled to adopt certain methods and points of view in the interest of the community and the state. There is self-help and governmental interference. The inter-relationships of these personal and external forces constitute one of the most important and difficult questions concerned with farming and politics. They are questions of adjustment between the self-help and the state-help forces as expressed in the complex terms of present society. One force works from the inside outward, the other from the outside inward. Both are essential. I propose to examine some phases of these questions that are suggested by current movements and discussions.

Governmental interference with persons is of two kinds,— that which concerns the larger relation of each man to society in general, and that which applies directly to the particular occupation, trade or profession. In the first division are questions of taxation, tariffs, conduct and the fundamental laws. In the second division, so far as the farmer is concerned, are questions of agricultural education, regulation of diseases of live-stock and crops, the establishment of institutions or agencies—as state stud farms—that provide him with direct facilities for improving his methods, and special applications of general laws. It is only some phases of the second class of governmental interference that I propose to discuss in this volume, although I suspect that some of the general laws need rather radical overhauling, if the farmer is to be dealt with in perfect justice. My present theme is this: What is it wise and legitimate for governments to do in aid of the farmer, and how, in general, may it be accomplished? In developing the discussion, all I hope to do is to establish a point of view. The instances and examples are cited as illustrations of what I mean to teach, rather than as specific problems that I would here work out.

I
The Agricultural Shift

THE condition of agriculture and country life in North America has been modified by three great shifts,—the shift in geographical location, in methods of practice, and in institutions. Some of the more obvious features of these shifts we must briefly examine.

THE GEOGRAPHICAL SHIFT IN RURAL OCCUPATIONS

Both the necessity for governmental interference and the nature of the interference are conditioned on the status of the rural industries at any particular time. We are all aware that we live in a time of great shift. The center of population is moving westward, and there is movement from country to city. Remote lands in the East tend to lose in population, and remote lands in the West tend to gain in population. There are still great areas of new development and consequently of unstable conditions. The geography of markets has undergone great change. The great shift in the East has come about very largely as a result of free trade with the West.

The popular mind has pictured a great decline in eastern agriculture and a corresponding increase in efficiency of western agriculture. This opinion is founded in part on statistics and in part on the larger base on which western agriculture is often conducted. This is often more apparent than real. We may follow some of the statistical comparisons between New York and some of the corn-belt states by way of illustration, and later we shall endeavor to determine the significance of some of the changes. Comparison by states is, of course, always indecisive and often very fallacious, because the state unit is not uniform in size, population or general condition; but we have no better way at present of making rapid contrasts.

In 1850, 1860 and 1870 New York held first

place in the value of farm property. In 1880 it lost first place to Ohio; in 1890 it took third place, being exceeded by Illinois and Ohio; in 1900 it took fourth place, being exceeded by Illinois, Iowa and Ohio. In population there has been a marked decline in rural communities. According to the figures of Rossiter, in 1850 five rural counties in New York showed a decrease in population; in 1860, nine; in 1870, nineteen; in 1880, eight; in 1890, twenty-three; in 1900, twenty-two; in 1905 (state census), twenty-one. It appears that forty-three counties have shown a decrease in population at some period during the past century. Twenty-eight counties, or one-half those outside the metropolitan districts, have a smaller population today than they have had at some previous time, and these counties represent nearly one-half the entire area of the state. There has been a decline under the maximum of more than eighty thousand persons in the rural counties of the state. Rural communities in some parts of New England have less population today than they had one hundred years ago. This decline seems to be expressed (1) in migration of population to cities and to other regions; (2) in lower birth-rate.

In New York from 1880 to 1900 there was an annual decrease in value of farm property, if the census figures are comparable, of seven and one-third millions of dollars. For the same period there was an annual decrease in the value of land and improvements of nearly eight and one-half millions of dollars. Similar apparent depreciation occurred in other eastern states. In Ohio, for example, the shrinkage of land values from 1880 to 1900 amounted to more than sixty millions of dollars.

The apparent reaction.

It should be said, before passing this subject, that it may be a question whether the census figures of the different years are in all respects comparable. Conditions of money and of values are not the same at any two twenty-year periods. In 1880, we may not yet have passed altogether the inflated values of the war period. These census figures are now old and great changes have taken place in the seven or eight years since the more recent ones were made. Some of these changes seem to be indicated in the most recent figures. A current discussion of "changes in farm values" published by the United States Department of Agriculture and covering the years 1900 to 1905, makes a very different showing from those that we have been in the habit of quoting. These figures of the Department of Agriculture are estimates and computations, and I do not know whether they or the census figures more accurately represent the exact status of agricultural conditions. Even for the census year 1900, the differences in values as reported by the census and as computed by the Department of Agriculture amounted for New York state to nearly ninety-nine millions of dollars for the value of land and improvements, including buildings. The computations of the Department as between the years 1900 and 1905 show a gain in similar values for the state of New York of more than one hundred and eighty millions of dollars. In more specific categories, the following figures from the same source show that there is a decided increase in farm values and, therefore, presumably in farm efficiency. The values of "medium farms" per acre for the years 1900 and 1905 in New York in the different classes of farming are as follows:

	1900	1905
Hay and grain farms	$40 29	$44 38
Live-stock	33 83	37 94
Dairying	46 81	58 86
Fruit	70 87	84 46
Vegetables	69 98	81 91
General farming	38 98	44 00

The percentage increase of real estate value of such farms in the state for the years 1900 to 1905 are represented by the following figures:

	1900–1905 Per cent
All medium farms	18.3
Hay and grain	10.2
Live-stock	12.1
Dairy farms	25.7
Fruit	19.2
Vegetables	17.0
General farming	12.9

These various figures are given here merely to illustrate the fact that the geographical base of agriculture has changed and that it may be expected still further to change; and that a reaction is likely to follow a great shift. Any shift of considerable area is likely to affect some localities disadvantageously.

THE SHIFT IN AGRICULTURAL METHOD

Farming exhibits the remarkable changes that have taken place in the last fifty years in the modes of doing work. The plow is still called a plow and for the most part it is yet drawn by horses (as it will continue to be drawn); but even the plow is a very different implement from its predecessor of a generation ago. Few implements are more perfect than the present-day plow in the application of mechanical principles and in workmanship. The slight variations in the slope and shape of the moldboard and in the construction of other parts, produce marked results in the effect on the physical condition of the plowed land. The chilled steel construction has produced an implement of new adaptabilities. The plow has come to be a construction of lightness, grace and beauty, and of great effectiveness.

Most agricultural tools have shown a similar evolution. Even the common fork has undergone marked change. Light hand tools are of many new kinds and forms. But it is in the range of machinery that the change appeals most to the imagination. To be sure, the great development of farm machinery has been mostly for the easy conditions of level-area farming and for the more wholesale operations, but the development has been marvelous nevertheless. The mere mention of the names of some of the farm machines will recall how great the change has been: from the sickle to the cradle, to the reaper and self-binder; from the scythe to the mowing-machine; the hay-rakes and hay-loaders; all the hay forks and stackers; the corn-harvester; the great separators or threshers; the sowers, planters and transplanters; manure-spreaders; the grinders and feed-mills; power ditching-machines; the spraying machines; the new kinds of vehicles; all the multitude of special milk-manipulating, butter-working and cheese-making devices; the adaptation of steam, gasolene and even electric power; and the marvelous range and beauty of tilling implements and machines. Within fifty years, the cost of producing a bushel of corn has been reduced by two-thirds; a ton of hay by nearly as much; and of other products in similar proportion.

The change is probably even more remarkable in the farmer's attitude toward the reasons that underlie his work, although this shift does not appeal so much to the popular imagination. His attitude toward soil fertility has undergone a complete change; so has his attitude toward the feeding of animals and the treatment of their ailments; so has it toward diseases of plants and toward the insects. He speaks a new language. Even when the old farm seems to show no visible change in external matters, the farmer himself cannot avoid attacking his problems in a new way. Butter-fat is a reality. There are new crops on his land—alfalfa, cowpeas, crimson clover, macaroni wheat. If he lives in the northeastern market-milk section, he has seen the red and brindle cow change to black and white; he has developed the winter production of milk and has made the silo a part of his farm scheme. He has a new conception of cleanliness, as a result of the studies in bacteria. He has a rational outlook on potato blight and oat smut and codlin-moth. He has respect for ideas in print, because the ideas are worthy of respect. All this changes his methods of work.

With all these great shifts in the methods of farming, it is natural to expect unequal shifts in effectiveness of the business, for some persons react responsively to such changes and others do not. The least adaptable persons find their lot harder by competition with the others. Profound changes have resulted in the whole attitude of the man toward his business.

These shifts in mental attitude are largely the direct result of the colleges and experiment stations and bureaus devoted to agriculture; and herein is the one great aid that society has rendered to the countryman.

THE SHIFT IN RURAL INSTITUTIONS

It requires no imagination to see that rural life is, in some respects, in a state of arrested development, as compared with the cities and towns. The native institutions have been copied in cities and greatly extended; the rural population now looks beyond its own institutions to those of the city,— to the city school, the city church, the city library, the city stores, the city amusements. The great constructive movements of the day have passed the country by. The nativeness of rural institutions has been allowed to die out, and the country has been left socially sterilized. Centers of interest are elsewhere. In many regions, the farmer will talk politics, war and city questions and anything else rather than farming. I would not have my reader feel, however, that this is peculiar to these later days. I well remember vehement discussions whether the pen is mightier than the sword, but I never heard a debate on the plow, which is really mightier than either.

Many great economic and social changes have directed the attention of all the people cityward. Canals, railroads, telegraphs, postal routes have drained the country into the city. Wealth has been piled up at the terminals, which are the trading places, until society has become ganglionic in its organization. Banking systems take the money from the hands of those who earn it, and put it into the hands of those who trade with it. The earnings tend to leave the place of their origin to build up remote or aggregated interests. The organizations that control farmers by controlling their products are in the cities. The tariff-for-protection system has fostered this general aggregational movement. It has tended to the concentration of wealth. If it has aided the farmer, it is because it has aided some one else first and more.

We have been living in an epoch of city development, with no adequate means of redistributing or returning the energy to the regions of its origin. It has been a process of dump. We are now, however, at the beginning of a new species of rural drainage, consequent on the wide extension of highway building, of trolly lines, of rural free deliveries, of telephones and other local-centering agencies. In other words, we are now entering the epoch of the small city; into these cities the surrounding country now will drain. This will develop new centers of influence, with a consequent shift of the social equilibrium. This condition is being aided from the city itself in the rapid growth of suburbanism. These new conditions constitute one step towards vitalizing the open country, but of themselves they will not reach the open country effectively.

Among existing rural institutions, the church and the school should have most influence; yet the rural church is largely inert or lost, and the school is in a state of arrested development. Following a discussion of abandoned farms in one of the eastern states, a farmer's wife, long a teacher and leader, wrote me the following letter; this letter I publish because it is an expression of the way in which some of these questions appeal to an experienced person on the spot; how widely it applies I do not know.

"The neglected farms of L—— should not be charged against either indolence or agriculture, because the main business of the township, extending over a period of twenty years or more, has been a religious war. There are three churches in the Town Center, representing as many denominations. Sometimes one has flourished, sometimes another, but sentiment respecting all of them is ever active. I have known crops to be neglected, work delayed, families divided, while the combatants awaited the outcome of some petty squabble over church affairs. One not familiar with conditions can hardly imagine the littleness of the superannuated gospel-splitters who are often sent to such outlying parishes. The war has been a continuous one for years. One pastor after another departed, in dudgeon, to have the combat renewed by the next one. This is all matter of public record, if any one wishes to inform himself. You may or may not have knowledge of this, but it is not a local phase. It is my firm belief that this is the situation in other 'abandoned farm' districts if the truth were known.

"Much the same thing can be said of the schools. You will not agree with me, but after spending some years in L———, and having been reared in a family of teachers, I am convinced that the schools also are often 'abandoned.' School officers may be too busy working their own farms and interests. They often have not the first idea of broad policy or even worthy public duty. I am not sure but that the farm might produce men of better caliber if we could ever get rid of the theory that grammar is more important than death or taxes. No close study of the town of L——— is needed to note the marked mental degeneracy of its younger members against the earlier men.

"And Agriculture? Well, no one ever built a church or school for agriculture. The silence of nature is a poor competitor amid the loud acclaim of decimal and dogma. After I was married, many of my relatives and friends in L———, observing my husband's methods, came to him for counsel. He hammered one relative into going every day (several miles) to the Exter creamery, even with empty cans, if need be, to induce others to join him. He placed in the town the first pure-bred bull, and now several others have followed. He sold cows at reduced prices to get herds started. He induced another relative, who lives on the homestead and is making money, to take his grades and rear them as an object lesson. He went to the Exters and pointed out the advantages of a milk-station in that section. 'Get something started,' he would say; 'it will gather strength as it goes.' Now they have the skimming-station, and, for the first time in years, L——— is looking men squarely in the face. The 'conversation' is changed. The Exters (splendid characters) have at last demonstrated that a little separator in a creamery is better than a big separator in a church.

"No greater error could be made, by men of intelligence, than to cry abandoned farms when abandoned brains is meant. L——— is a beautiful, fertile valley. The neglected farms are effect, not cause. The abandoned church and the abandoned-school-houses are standing directly in the way of rural progress and all the efforts of agricultural teachers cannot overcome the loud paternalism of these two powerful obsolete institutions that stand like feudal castles awaiting the dynamite of revelation.

"But, after all is told—mark this! Out of that 'God forsaken' town have come the best men, the most sterling characters (independent of their financial troubles) that I have ever known. The happiest years of my life were spent there. My little girl's fine father was born and reared there. The world has its compensations. I know and love these people, and the most impressive picture I will ever have of the great Field Shepherd with his people hovering about him, is outlined by a faded wall in that farming town."

THE "ABANDONED FARM" AS AN ILLUSTRATION OF THE AGRICULTURAL SHIFT

The process of the shift in occupancy of land has resulted in the desertion of some areas for customary agricultural uses. Homesteads have become unoccupied and the buildings have fallen into ruin. To this condition the name "abandoned farms" has been applied. The words are inexact, since the land is really not abandoned, but belongs to some one, and is still classed as property; but the term has become so well fixed in the public mind that it expresses phases of our contemporary agriculture better than any other phrase. I therefore use the term for convenience, to express the idea of the lessening present use of certain lands, and also because it enables me to arrive at my subject without tedious explanation.

In the discussion of abandoned farms, we have usually confused the economic and personal results. Our regret for the abandonment of these farms is, in fact, largely sentimental. We are thinking of the human lives that have been lived in the stark old houses, where the wind blows through the creaky roofs and into the openings that once were windows and doors. We are impressed by the ancient orchards, with bleaching limbs and rotting trunks. Generations ago, perhaps, the good-man and his wife hewed the farm from the wilderness, erected the laborious buildings, and planted the cider-apple trees between the rocks. A little brood was reared. One by one the children went out over the hills and made homes for themselves; but one of them remained on the old farm, cherishing the old rocks and living content under the old roof-tree. The old place became more hallowed with the years. Even the decay of the old house passed unnoticed. What was the neglect and dilapidation to the visitor from the city, was the object of veneration and sacred memories to the owner. But finally no youngster clung to the homestead. One went to the city store, another on the railroad, another took to sea, and another went west. Age overtook the old folks. The bushes encroached on the back lot. The stone walls fell to ruin. The kitchen roof tilted and fell in. The old folks died. The house was closed. The stripling forest overran the orchard and the garden. The tansy now haunts the old dooryard.

I have no desire to analyze at this time the causes of the so-called abandonment of farms, or to make any close study of the results; I wish only to call attention to some of the grosser features of the movement, with the purpose of establishing a point of view for my reader on some of the coming relations between the state and the farmer.

The "abandoned farm" question is supposed to be an eastern question, and in a way it is so: that is, relatively difficult lands were settled in the East because other lands were not available; the adjustment to changing conditions comes first, of course, in the older communities, and the older the community the slower the adjustment is likely to be. Similar questions are pressing agriculture in all parts of the country, however, so that the eastern question is only a phase of continental or even world-wide questions. The farming in the great agricultural West has been easy, and it has therefore risen with phenomenal rapidity. The greatest skill in the end develops under the greatest difficulty. The West has set the East many good examples of agricultural practice. It will not be surprising if the East works out some of the difficult social problems, and makes a close adjustment of agricultural practice to local conditions.

The illusion of old buildings.

A common measure of the supposed decline of farming is the fact that many farms can now be purchased for less than the buildings cost. This statement of itself does not appeal to me as having any special significance. A property is likely to sell for what it is worth, and this worth depends on its effectiveness as an economic unit or enterprise. Most of the buildings on these farms were erected a generation or more ago, when the ideas of farming were radically different from those of the present day. It is doubtful whether most of these buildings were ever really effective even for the old kind of agriculture. At all events, few of them are adapted to the business that we must now conduct on the land. Many a farm would really be worth more with the buildings off than with them on, for they would not then stand in the way of actual betterment. Buildings are not permanent attachments to land and should not be so regarded. A countryman is always impressed, when he goes to the great cities, with the fact that buildings still in good state of preservation are torn down to make place for new ones. These demolished buildings may not even be very old, but they are ineffective for present-day business and it is unprofitable to keep them. The coming business of farming will demand a wholly new type of building in order to make the property effective, and we must overcome our habit of harking back to the time when the present buildings were erected. Barns and other business buildings that were erected fifty or sixty years ago should owe the farm nothing by this time. My reader must realize the fact that we are beginning a new agriculture, not continuing an old one.

We must be careful, also, not to be misled merely by the appearance of farm property. The mere abandonment of farm buildings may or may not be a cause of apprehension and regret. Buildings may be abandoned because two or more properties have been combined into one and not so many buildings are now needed; or because the farmer has moved from an old building into a new and better one. In many parts of the East, the buildings are no doubt too many and the farm properties too small for the greatest effectiveness. These properties were laid out or divided at a time when the great West was unknown and when these eastern lands grew the grain and other tilled crops for the large markets. Some of them were probably laid out in their present form in war time, when conditions were wholly abnormal. Many of the buildings were erected when lumber and other materials were cheap and when the comforts and facilities now placed in barns and residences were unknown. Moreover, deserted farm buildings are likely to stand until they fall down. In cities, land and location are valuable, and old buildings are torn down to make room for the new structures. Therefore, the country contrasts strongly with the city in respect to its buildings. The staring and windowless farmhouses appeal to the imagination of the town visitor, and he accepts them at once as evidences of failure and decline.

In order to determine the significance of deserted farmhouses, an inquiry has been made in one of the townships in a so-called abandoned farm region. Every deserted farmhouse in the township was recorded. Conditions there are as bad as anywhere in that region; yet by count, there are only forty-five vacant farmhouses in the township, the area of which is more than forty-five square miles. One might draw the conclusion at once that there are forty-five abandoned farms in the township. It is doubtful, however, whether there is a single really abandoned farm in this area. It

Deserted Fields

is true that there are many fields on the higher farms that are not used except for hay and pasture, and some of them are not even used for these purposes. Practically all these so-called abandoned farms are either owned or rented by near-by farmers and have really become a part of the adjacent farm. The house is unoccupied for the simple reason that the farmer needs but one house. A few vacant houses have been deserted by families who have lost their homes on mortgage, but apparently not primarily from fault of the land. Many others have been sold because of discontent on the part of the owners, who wished to try their fortunes elsewhere. In some cases the owner has died and the house is unoccupied because the estate has not yet been settled. A few more are vacant because tenants cannot be secured, and the farm is rented to whomever is willing to take it on shares.

Old fields.

Similar remarks may be made with respect to many of the apparently abandoned fields.

Because of inability to secure labor, the fence-rows and fences are often not as clean as formerly, and the roadsides have a shabby appearance. Fields are often grown to weeds; yet these fields may be only resting until the owner finds time to put them into crop, or they may be used for light pasture, or they may be in the process of returning to forest. Of course, they are relatively ineffective pieces of property, but the conclusion must not be reached, because they are unkempt and not in use at the time, that they are abandoned or that the owner considers that he is obliged to desert them.

The significance of the general situation.

It is unquestionably true that there is lessening utility of some of our farming lands. In the face of this fact, however, three other facts stand out prominently: (1) Markets are as good as ever, for there is no decline in the purchasing power of the people (rather there is a reverse tendency); (2) the land is still productive, notwithstanding a popular impression to the contrary; (3) good farmers are better off today than they ever were before.

We have heard so much about the abandonment of farms that we are likely to think that it measures a lessening efficiency of agriculture. We must not be misled, however, by surface indications. We are now in the midst of a process of the survival of the fit. Two opposite movements are very apparent in the agriculture of the time: certain farmers are increasing in prosperity, and certain other farmers are decreasing in prosperity. The former class is gradually occupying the land and extending its power and influence.

The older farming was practically a completely self-regulating business, comprising not only the raising of food and of material for clothing, but also the preparation and manufacture of these products. The farmer depended on himself, having little necessity for neighbors or for association with other crafts. In the breaking up of the old stratification under the development of manufacture and transportation and the consequent recrystallizing of society, the old line fence still remained; persons clung to the farm as if it were a divinely ordained and indivisible unit.

We are now approaching a time when the traditional boundaries must often be disregarded. The old farms are largely social or traditional rather than economic units. Because a certain eighty acres is enclosed with one kind of fence and assessed to one man does not signify that it has the proper combination of conditions to make a good farm.

We must consider that the agriculture of the eastern states is now changing rapidly. It has passed through several epochs. The possibilities of agriculture in the East lie largely in a new adaptation to conditions, and in its diversification. This diversification is already a feature of the East. It is significant to note that while New York, for example, ranks fourth in value of farm property, it ranks as low as seventeenth in farm acreage, showing that the yield per acre is far greater than in many of the competing states. In the total value of farm products, New York is exceeded by Iowa, Illinois and Ohio. In the value of farm crops, in 1899, it held fifth place, being exceeded by Illinois, Iowa, Texas and Ohio. Considered with reference to the value of farm products per acre, it leads the states in this list, the figures being New York, $15.73 per acre; Ohio, $13.36; Illinois, $12.48; Texas, $12.25; Iowa, $12.22; and New York is exceeded by New Jersey and most of the New England states. Considering the fact that New York state is one of the largest states east of the Mississippi, this condition also indicates that New York is internally less developed than some of its competing states. Illinois ranks first in value of farm property and first in available farm acreage; Iowa ranks second in the value of farm property and second in available acreage; Ohio ranks third in value of farm property and third in available acreage; New York ranks fourth in value of farm property and seventeenth in available acreage. The above statements suggest the reverse of decadence in eastern agriculture, whatever may be the statistics that express changing values or whatever may be the popular fancy to the contrary.

A further evidence of the great diversification of agricultural enterprises in New York, as a representative of eastern conditions, is shown by the fact that in a list now before me of twenty-two leading products of this latitude, New York stands first in the production of eleven of them, whereas no other state ranks first in more than two or three of them. While the agriculture of the state in general shows a decline as measured by the census figures, the main lines of special development are in a condition of increased vigor and effectiveness; and this remark may be extended to other eastern states. The wonder is not that certain lands are returning to forest, but that, in all this shift and the rapid development of the West, the state has been able to hold the position that it still maintains.

This rapidly moving readjustment and diversification will produce fundamental changes in the mode of farming and in the economic, social and political outlook of the people. In the mode of farming, it will force new business organization; and when new acres cannot be had, the old acres will be doubled by using them to greater depths. In very many ways, the shift is now demanding a new kind of study of agricultural questions. This redirection of agriculture is bound to come in every state; and we should meet it hopefully.

Nor would I have my reader feel that this readjustment is all in the future. It is proceeding at the present time, and with greater momentum and effectiveness than many of us, I suspect, are aware. After many years of touch with the problem and with the men who are capable of judging it, I am impressed that the persons who are most alarmed are those confined largely to offices and who are given to the study of statistics.

The situation with individual farms.

A discussion of statistical generalities does not exhibit the status of the individual farmer nor give us specific reasons for the decline of profitableness in farming. Every farm is a problem by itself and what may have been responsible for the defeat of one farmer may not have been the cause of the embarrassment of his neighbor. Some of the decline no doubt lies directly with the man, quite independently of the land: it is psychological and perhaps even hereditary, and in its community aspects it is social; but these phases I am not now prepared to discuss.

The larger number of the farms of apparently declining efficiency are in the hill regions. In New York, many of them are on soils of the volusia series, particularly on the volusia silt loam. This soil is of low humus content, usually with a high and compact subsoil, and limited root area. Many of these farms are unsuccessful in part because of their climate. They are elevated. It is often impossible to grow with profit the common varieties of corn and even of other grain. Sometimes the difficulty lies in their remoteness and the cost of transportation, together with the poor schools and social disadvantages that are a part of such isolation. Usually these hill lands are expensive to work, and they do not lend themselves well to open tillage. Very frequently they suffer for lack of under-drainage. If the elevation is too high to grow good wheat it may also be too high for good clover, since clover is usually seeded with the wheat.

These high and rough lands are not so frequently plowed as lower and flat lands and, therefore, they are not cleaned, do not receive the benefit of rotation, and they are likely gradually to deteriorate in physical condition.

There has also been great change in market demands. Beef-raising has gone out of the East. It was a simple thing to grow the beef and to raise the milk in the old time, but it requires skill to grow and market a modern steer and to tend a modern dairy herd. With relatively few cattle, there is insufficient enrichment of land. The farmer on these hills is likely to practice direct sales; that is, he sells his timothy hay and other products direct, removing thereby a large amount of fertilizing value and saving nothing of the crop except the roots and stubble to return to the land. This primitive mode of general farming allows a man to make a profit only on a single sale. The manufacturer tries to turn his property over more than once, each time expecting to realize a profit. When the farmer is able to market his forage largely in the shape of animal produce, he will not only save fertility but should make a profit on both the crop and the animal. The selling of baled hay rather than pork and beef and milk and eggs, cannot be expected to yield much profit or satisfaction to the average farmer or to keep his land in living condition. Taking it by and large, no agriculture is successful without an animal husbandry.

The popular mind pictures these so-called abandoned lands as exhausted in their plant-food, but this is probably not often the case. Very many of them are potentially as productive as ever, but they are run down; yet even at their best they might not be able to satisfy a man who lives in the twentieth century. Human wants have increased. What would have made a good and comfortable living seventy-five or one hundred years ago, would not support a man in the way in which he ought to live today, nor would it attract his boys to remain on the land.

Lack of adaptation.

All these and other causes of the decline of individual farms can be expressed as a lack of adaptation to the natural surrounding conditions. Good agriculture is the perfect adjustment of the methods of the farmer to the particular region and circumstances, thus making all effort count and eliminating waste. This is why some of the European farming is so much better than our own. In the end, therefore, good farming is not a question of West or East. One often finds excellent farming in what are generally considered to be poor agricultural regions.

It is a biological fact that animals and plants cannot thrive unless they are well adapted to the conditions in which they live; and if they are wholly unadapted, they perish.

Now, farming in this country is not yet adapted to the natural conditions of soil and climate and market and other environmental factors. In fact, we really do not yet know what the soil factors are, if, indeed, we know to any degree of accuracy what any local factors are. If some of our eastern farms have changed from corn and wheat to hay, and if they have not prospered under this change, then it follows that they have not yet found their proper adaptation. It is not at all strange that this adaptation is lacking, since there has been no means of putting the farmer into touch with his own problem. Not one of the older farmers was adapted to his environment by the church or the school or by any other educational or social agency. If he is now adapted to the conditions in which he lives, it is because of some accident of heredity or circumstance, or because of his native wit. We can never adapt the business of the farm to its conditions until we understand thoroughly all the problems involved, and there has been no serious effort to understand these particular problems until within very recent time.

Much has been said about the disadvantage of the eastern farms in competing with the western farms. I am convinced that they often suffer quite as much by competing with each other or with regions close at hand. In a thirty-mile drive, I traveled a flat country where oats were a good crop and harvested by machinery and drawn from the fields in high-piled racks; I also traversed a country of high and steep hills in which oats were a poor crop and not harvested by machinery and were hauled from the declivities in small loads. It was evident that the latter region could not compete in the raising of oats with the former, although they were less than twenty miles apart. The one region seemed to be well adapted to oats and the other, at least on the hillsides, was not a profitable oat country. In other words, the farmers on the hills had not adapted their farming to the hills. I suspect that a bushel of oats cost them at least 50 per cent more than it cost the men at the other end of the county. Yet, I think that there is a way of profitably farming such hills: many men have proved it.

Point of view as to remedies.

While I am convinced that the general condition of eastern agriculture is prosperous and hopeful, we all know that there are very great problems and that some regions are much more disadvantaged than others. If we are to discuss remedies we must first of all establish a point of view.

We must first disabuse our minds of all prejudgments and consider the conditions as they actually exist and in their relations to the general progress of the race. Our outlook must be forward rather than backward. We must overcome the influences of many phrases and trite statements that have long been public property. It is said that the farms are the bulwark of the nation. Like all trite sayings, this is both true and false. We need the conservative element of the farm, that has its feet planted directly on the verities of the earth.

But we must remember that poor lands usually raise poor people. I do not conceive it to be necessary that all the lands in any commonwealth should support farm families in the sense in which we have understood it in the past. It is much better for the commonwealth, both from the economic and social points of view, that many of the lands should be devoted to forests or even allowed to run wild than that they produce people that are only half alive. I should want to keep the conservatism of the agricultural peoples, but I should want this conservatism to be constructive and progressive.

I am not ready to admit that the traditional "independent" farm family on 80 or 100 acres is always necessarily essential, as we have been taught, to the maintenance of democratic institutions or to the best development of agriculture. The size of holdings and the relation of the family to the land, are likely to change radically in many regions, and we must be prepared to accept the fact. The American has a traditional fear of large estates, but such estates are bound to come in some of the remoter regions. We should now be sufficiently established in democracy to have forgotten our early alarm at such estates. Very likely we shall repeat to some extent the experience of Germany and other countries, where leadership of large agricultural estates has contributed to welfare.

In the discussion of abandoned farms, I fear that we have been misled or even scared by a phrase. We have accepted the term "abandoned farms" as itself a statement of fact and have seemed to reason from it as if it presented a single condition of affairs. Our imagination has often outrun our reason. It is not so much a question of abandonment as of shifting occupancy and radically changed conditions. If these conditions had been expressed with equal emphasis by some other phrase, the discussion of the question might have taken a wholly different direction. Suppose, for example, that a part of the problem had been expressed in the term "farms becoming forested:" the least imaginative of

Not an Isolated Question

my readers will at once see that a wholly unlike line of thought might have evolved from the discussion and wholly different conclusions might have been reached. There is really no problem of abandoned farms as such. The so-called abandonment of farms does not represent one condition but many conditions; not one series of facts but many series of facts; not one forthcoming result but many results. The condition of agriculture, even though we admit it to be bad in many particulars, is not a cause for alarm, but is rather a reason for new and careful study. Nor does this condition affect agriculture alone; it is rather a problem of economic evolution, that concerns the organization of society, and consideration of it cannot be separated from the discussion of general welfare questions of the day.

Mere public propaganda cannot solve these questions of land occupancy. Associations and conventions cannot solve them. Importations of labor cannot solve them, much as it may help the individual farmer here and there. It is a debatable question whether we should try to restock many of the present farms merely by putting a foreign family on them. Perhaps the very reason why these farms are in the process of decline is that they are necessarily ineffective economic units and are not capable of being directed into a farm management that is adaptable to present conditions. Merely to put families back on many of these farms would be to continue the old order; and it is this old order that we need to modify or to outgrow.

Viewed as an economic question, the shifting of farm occupation should not disturb us more than other shifting of population. In the present day, some of the lands that are now "abandoned" would not have been settled. They would remain in timber; and now, by the inexorable power of economic forces, they are returning into woodland. Some of these farms ought to be abandoned to other uses. It is a misfortune for a man to be obliged to inherit one of them, and be sentenced for life to live on it. He would much better try to escape.

No mere treatment of symptoms can have much permanent effect on agricultural conditions. Many agricultural localities are making great effort to secure summer boarders. This may aid a certain class of persons; but as the summer boarder advances into the open country, agriculture is likely to recede. The solution of the problem is a long-time process. It is not merely adding fertilizer, nor killing daisies and paint-brush; it may not be even a question of making the farm more productive. The little-farm-well-tilled idea will not solve the problem. It must be a process of reorganization.

Let us bear in mind that the questions of ineffective farming are not new. Just now the emphasis seems to be placed on the so-called abandonment of farms, and on certain kinds of propaganda that promise to solve these difficulties. We have passed through many epochs or eras of wide-spread propaganda, in each one of which some one factor was supposed to afford the means of relieving agricultural distress. I remember that at one time the emphasis in agricultural discussion was placed largely on the farm mortgage, but we have learned that a mortgage on a farm is not inherently different from a mortgage on any other property. I recall very well when the era of compounded fertilizers was at its height: all one had to do was to have the soil and plant analyzed to determine the deficiencies, and then to prepare a medicine to cure the disorder. I remember the advent of farm machinery, which was supposed to be able to solve the farmer's difficulties. I saw the beginning of spraying for insects and plant diseases, and it was figured up for us what losses we suffer from bugs that prey on our crops; it has cost us more to fight bugs than to fight Indians, counting the value of crops that they destroy; spraying would provide a remedy, and yet bugs are still with us. At one time the emphasis was placed on under-drainage, and we need a recrudescence of this teaching. In parts of the great West, the emphasis is naturally placed on irrigation. We have looked to the rural free deliveries of mail as one of the great means of alleviating agricultural isolation and failure. The good-roads people have been sure that the lack of traversable highways is the cause of the so-called agricultural decline. Lately, various kinds of extension work have been strongly in the public mind. We are just now in the era of soil surveys and other soil studies. We are beginning to talk in a new way about the old and yet unknown subject of farm management. We are talking freely of social questions, without knowing just what they are.

Every one of these epochs has placed us on a higher plane, and yet we have never heard more about agricultural decline than within the past ten and twenty years, notwithstanding that this is the very time when the agricultural colleges and experiment stations and governmental departments have been expanding knowledge and extending their influence. The fact is, that all these agencies relieve first the good farmers. They aid those who reach out for new knowledge and for better things. The man who is strongly disadvantaged by natural location or other circumstances, is the last to avail himself of all these privileges. We have learned that it is not sufficient merely to start good movements, but that we must have some active means of reaching the last man on the last farm, so long as he lives there. This is by no means a missionary work; it is rather a duty that the state owes to its citizens, to provide those persons in difficult positions with the best possible means of making their property thoroughly serviceable. It becomes in the end, therefore, a personal question as to how information and education can be taken to the farms in such a way that the farming shall profitably adapt itself to its environments. The failure of a great many farmers may be less a fault of their own than a disadvantage of the conditions in which they find themselves.

It is fairly incumbent on the state organization to provide effective means of increasing the satisfaction and profit of farming in the less-fortunate areas as well as in the favorable ones, both as an agency of developing citizenship and as a means of increasing the wealth of the state. The state cannot delegate this work, nor can it escape the responsibility of it. It is primarily an internal question. The questions must be attacked just where they exist, and with the sole purpose of solving them for the good of those who meet them.

The outlook for the hills and remote lands.

Wherever farming is not now profitable, a special effort should be made to readjust the handling of the lands to the conditions of climate, soil topography, markets and the like. Any one who has traveled much in the northern states will have noticed the superior quality of the tree growth and the grass cover in that region. Of course, the unproductive areas, whether on hills or plains, present very many conditions and they may be adaptable to many kinds of agriculture; but in the particular type of hill land and remote land which is now most in the public mind, I look for the development of at least three strong forms of farming:

(1) Fruit-growing for export. We have developed great skill in the methods of car-

ing for orchards on the relatively level lands of the special fruit sections, but we have given very little attention to the growing of first quality apples in the more hilly regions. In such regions we cannot practice the type of clean tillage that we advise for other lands. Some relatively simple and inexpensive type of farm management must be applied to them. There is every reason to think that large areas in the East that are now practically unknown to fruit may grow a grade of apples that will be in great demand in the foreign trade. The state can well afford to undertake some large demonstrations in the growing of such orchards.

(2) A revival of the animal industries and the extension of dairying. With the continued development of great city markets, the dairy industry must grow. Many of the hill and outlying lands are no doubt admirably adapted to pasturage and forage crops for cattle and sheep and swine; but the livestock interest, aside from dairying and poultry-raising, is altogether too small in

the East. The eastern states should now be making inquiries into the condition of the animal husbandries within their borders.

(3) The growing of forests. It is to the forest crop that vast areas of the roughest, highest and most unproductive lands of the East are best adapted. As near as I can determine nearly one-third of New York, for example, is in woodland. In some counties, even outside the Adirondack reservation, two-fifths of the land is reported to be in wood-lots. This is a greater area than is devoted to any other crop, and it probably yields less profit per acre; yet in the census year New York led all the states of the union in the value of farm-forest products.

As a people, we must re-orient ourselves to the subject of forests. The forest is, or ought to be, considered as a crop. Natural forests are not necessarily the best forests, so far as the production of timber is concerned. Nearly all natural forests abound in unproductive areas, and in trees of very slight commercial value, which are as much weeds in the forest as Canada thistles are weeds in the corn-field. Man can produce a better commercial forest than Nature usually does.

These forests may well belong to the people. Schools and towns could be supported by the proceeds of good community forests, at the same time that water-supplies could be conserved, wild animals protected, and the beauty and respectability of the country enhanced. When this time begins to come, the commonwealths that have rough lands may consider themselves to be fortunate. The town, county or state could well afford to buy some of these lands and devote them to forests. The United States government is well begun on this process, and this is right; but it is also necessary that the states and communities themselves acquire forests in order to maintain their institutions and to develop local enterprise, the importance of which I shall try to develop in the second part of this book.

II
Society and the Farmer

I HAVE now tried to give my reader a picture of some of the conditions confronting the strictly rural society. Even this brief sketch is sufficient, I think, to suggest that the problems of this society are not agricultural problems alone, nor even rural problems alone. All society must interest itself in them. In particular, must the agents of society—the various organisms or departments of states and communities—extend their constructive work directly to the open country, not only that the interests of the open country may be advanced but that the welfare of society itself may be safeguarded.

I propose now to bring briefly and rapidly before the reader a few of the ways in which society, or the state, may exercise itself in this direction to advantage. I have in mind society in general,—that is, all men and women consciously or unconsciously working together for a result,—and not alone the mere formal organization of government. I mean to express some of the mutual obligations of the state and the farmer.

THE PROBLEM

We must never overlook the importance of these farm producers to society, nor forget that they deserve as much from society as any other persons. These persons are not much in evidence. This is important: they are not working for honor or acclaim. They are remote. This also is important: they are near the sources.

Saving our resources.

The memorable Conference of the Governors has left us with a new appreciation of the importance of our natural resources and the necessity of saving them. Much was said about the development of water-power, the preventing of land erosion, the importance of governmental regulation of forests. A number of the governors declared that they would appoint forest commissions on their return: this may be of value, but it is not likely to accomplish much, as commissions go. The man who stands at the sources, is the one on whom we must in the end depend for the work of preservation. The instincts of the settled farmer are all for preservation and betterment, not for exploitation or for sales of stocks: he is the natural conservator of the native resources of the earth.

There are many persons who are waiting to know what forces the great Conference will set in motion to reach and quicken the man at the sources; that is, how we are to get to the real bottom of the question. It was most interesting to follow the discussions on the means of developing water-power: the Mississippi, Niagara, and other great streams were mentioned. This development, of course, is necessary. But rivers are not born as rivers. They originate from a little lake in the mountains, and a rill in a forest, and a spring in the pasture lot. To a great extent, they originate or are supplied from sources on some man's land. This man has the first use of the water. Every farm supplies something to the rivers. Many of them supply living lakes and streams. There are more than five millions of farms in the United States. Every good farm will in time have its own mechanical power. Much of it will be water-power. When the farmer developes his water-power, he will also protect his stream or spring. It is more important that we develop small power on a million farms than that we organize power companies or harness Niagara.

Our natural resources are of three kinds: Those of the mining order, the supply of which we can prolong only by saving; those growing directly or indirectly out of the earth and sea, as all forests and other crops and all animals, the supply of which may not only be conserved but may be greatly increased; the streams and lakes, the control of which depends very directly on the crop-cover of the earth.

In the last analysis, the utilization of the powers of the earth depends on the man who raises the crops, whether of forests or cotton or wheat. The solution of the problem is to reach this man. This man is coming to a new sense of his responsibilities. We often say that the farmer feeds all the people. He must do more than this: he must leave his part of the earth's surface in more productive condition than when he received it. This he will accomplish by a better understanding of the powers of the soil and the means of conserving them, for every well-managed soil should grow richer rather than poorer; and, speaking broadly, the farm should have within itself the power of perpetuating itself. The enrichment of land by the mere purchase of mined fertilizers— which is transportation, or the exploitation of one place for the benefit of another,—will not accomplish this. Every young man going on the old farm should feel that he has practically a new farm to begin on; and every good farm should pay for itself, buildings and all, in every generation of men. A farm youth, as well as any other youth, should be able to start anew, if he wants to, even though he does not go west.

It is not only important to farming, but absolutely essential to the nation, that the man at the sources be reached. The farther removed the man, the nearer the sources he is likely to be, and the greater may be the necessity of reaching him. We have made only the merest beginning toward reaching him. We must not overlook any man.

Government can go into farming,—that is, into forest-farming—on its own account, and this it must do. But the one great thing that government can do for the man on the land, that it does not do for all men, is to increase his sense of responsibility to the land and to give him power to use the land. This is education by means of agriculture,—using the word agriculture broadly for man's occupational contact with the surface of the earth. This is the real solution of the problem of the saving and increasing of our natural resources. This lies beyond and behind all commissions and conventions. Perhaps the commissions and conventions will help to bring this about more speedily.

The social questions.

Country affairs must be redirected. The problem is chiefly social. Good farmers are making the farms pay. The financial part of the business is improving. The community feeling, however, seems to be dormant, if, in fact, not actually perishing in many places.

We need to give as much attention to the social welfare of the rural country as to similar questions of urban regions. Our studies of social questions have been confined very largely to congested populations, but these questions are just as many and just as important in communities of scattered homes as in cities and towns. Even the question of congestion of population is not a city problem alone. Part of the city population comes from the country and this movement may distress the country from under-populating it at the same time that it distresses the city from over-populating it. But even if city congestion were to come wholly from the city itself, it nevertheless would still affect the country, not only because some of the surplus might find its way into the country, but because all populations are inter-relating and inter-acting. While it is customary to divide human beings into city people and country people, all great human problems are fundamentally the same, differing chiefly in their phases and symptoms. There is a city phase and a country phase of every great question. The city phase has been studied with much care, and, therefore, we have come to think that social problems are city problems. But whatever vitally affects the city likewise in some degree affects the open country. One of the great needs of the time in social studies is that we discover the rural country.

There is a city phase or application and a rural application to all questions of education, truancy, public health, pauperism, immigration, charities, corrections, civic relations, labor, density of population, moral standards. We have made the serious mistake in treating some of these questions as separate problems for the city and the rural districts, largely, however, by disregarding the one. We are at this moment making the mistake of considering agricultural education as a thing apart, whereas it is only a phase of education in general and cannot be isolated without leading us into error.

If these statements are sound, it follows that the country should not be exploited in the interest of the city. The country must be developed for itself and out of itself. There must be a country social order, as there is a city social order. One might think, from many current discussions, that the country exists for the convenience and benefit of the city, providing occupation for those who have failed to attach themselves in cities and an asylum for the undesirables. To some persons, the country question seems to be only a congeries of isolated problems of needy families and of vicious communities; but these are not country questions more than city questions. In either case, they are but symptoms. Want must be relieved and vice must be controlled, wherever they are. No small part of the vice of the country districts is that which is forced out from the cities. The open country has problems enough of its own without being obliged to receive the over-plus from cities.

What I have in mind is far more than the mere relief of symptoms here and there. I want to see the development of a virile and effective rural society; and I know that such a society can come only as the result of forces arising directly out of the country, as a natural expression of the country itself, not as a reflection or transplanting of city institutions. The country must develop its own ideals and self-respect. My city friends, for example, are proposing ways whereby country people may have entertainment, but they make the fundamental error of fashioning their schemes on city ways. The real countryman does not think of theaters and recitals and receptions and functions in the way that the city man does, and it is not at all necessary that he should. On the contrary, it is very important that he should not. The countryman needs more social life; but his entertainment and contentment must come largely out of his occupation and his contact with nature, not from mere extraneous attractions. Herein lies the root of my concern in nature-study and nature-sympathy: the countryman must be able to interest himself spiritually in his native environment as his chief resource of power and happiness. Many a country family rusts and dies for want of a local stimulus.

Holding that it is fundamentally important to preserve and encourage originality in the open agricultural country itself, it will then be necessary to stimulate re-directive movements to prevent the country from tumbling headlong into the small city or town. The tendency of farmers to move into town is to be deprecated. It is not necessary to pause here to combat the prevalent but superficial notion that farmers would better live in hamlets because European farmers live in them, but only to say that the European custom is the result of historical and social conditions that do not obtain in this country, and to consider that Europe itself would undoubtedly be better off if it were possible for a different condition to obtain. We have no peasantry in the United States, at least not among the whites, and farmers are moving away from peasantry rather than towards it. The great social movement of the world is away from peasanthood. What it may be necessary to do to arrest the drainage to the small city, we shall presently consider.

The countryman.

The country problems must be approached sympathetically, from the standpoint of the countryman. The countryman is to live in the country, and to make it or mar it. Those who approach the subject with the idea that the countryman is unresponsive or incompetent, are really not in sight of the problem and would better let it alone. One who judges country life by city standards,—as many city persons do—would also better let the problem alone. Many of the criticisms of the personal appearance and habits of the farmer and the pictures of supposed unthrifty farm properties, only show that the author of them is looking at the question from the outside and at long range. It is quite as likely that it is the city man who is in need of help. It is the commonest thing for the onlooker to say that farming must be more "scientific." Of course this is true, but not in the way in which the onlooker commonly conceives it. It is the easiest thing to make the most stupid failures by merely appropriating the scientific facts and discoveries of the investigators; it is quite another thing to work these facts into a good system, but this is a matter of slow and laborious growth.

There are farmers and farmers. They are of all kinds and nationalities, and of all ranges of competency; but the good farmer is one of the most industrious, capable and steadfast of men, and he is likely to have a real sympathetic relationship to nature that stands him in good stead at all times. He does not need to have help or charities dispensed to him. But society needs to recognize him, and it is high time that the state should undertake positive constructive efforts that will allow him and aid him to express himself to the full. There is enough talent and ability in the rural country to have set the agricultural status far ahead of its present condition, if it only were called out and allowed to express itself. It lacks opportunity.

The best public opinion grows where men are most independent, where they are least tied commercially and personally to other men, where they may have an opinion without fearing to jeopardize their trade or their position. Commercial men and salaried men are tempted to be trimmers and compromisers. The method of "practical politics" is compromise. This independence can grow much better on land that one owns than in rented houses. It must proceed direct from cause to consequence, as a field of corn grows direct from seed to ear. The processes of the countryman are direct. They are not over-organized. There is fixity and directness of attention in the country. The man is not diverted by a thousand things. The city boy may know much more than the country boy, but a good deal of what he knows may not be worth knowing. When the man on the land is well educated in the terms of his environment, we shall have the kind of public opinion that stands. The roots of society are away back in the soil.

If society is under obligation to consider the farmer, it is equally true that the farmer bears an obligation to society. This the farmer is likely not to recognize. The means that will bring about the one, however, will necessarily bring about the other.

Rural needs.

Before we can intelligently discuss some of the remedies for the social ills of the open country, we must inventory the needs. These needs seem to fall chiefly into five great groups, which may be briefly stated:

(1) The need of greater technical knowledge of agriculture. This knowledge of discovery and teaching is being rapidly supplied by the experiment stations and colleges of agriculture. The knowledge that we already have is far in advance of the practice of it; this is necessarily true in any branch of human activity, but this

does not argue against the necessity of still more knowledge. As much as we have learned, all the great fundamental problems of rational agriculture are yet unsolved, and many of them are not even explored. Great as our lack is in these directions, it is perhaps even greater in the social and coöperative lines: the great country problems are now human rather than technically agricultural.

(2) Need of governmental protection, whereby the disabilities that are not a part of his business may be removed from the farmer. Governmental protection and control are least applicable and least effective in the farming country, and the farmer has more burdens to carry than those pertaining to the rearing of crops and animals and to the contest with climate and weather: some of these handicaps will be removed or their effects minimized in the future (page 81).

Corollary to this is the lack of any kind of organized supervision over country living. For example, there is no continuing oversight of public health in the farming country, except a more or less effective effort when communicable diseases break out; and the supervision even then is usually more in the interest of the city than of the country. We are much in need of health supervision directly from the country point of view. The lack of attention to health regulations is little less than appalling in its consequences. The physicians in the farming country are general practitioners, commonly out of close touch with specialists and experts. It is pitiable that so many of the good country population are lost from neglect, and antiquated treatment of disease. I have no means of knowing whether the country suffers more than the city in this particular regard; but well enforced sanitary regulations are powerful educators, and the country does not have the benefit of them to the same extent that the city has.

(3) Need of the coöperative spirit in business. The development of our rural country has proceeded mostly on the basis of isolated occupancy of land, with the strong individualism that goes with it. Definite coöperation has not

been necessary; and what has once become an established order soon becomes tradition.

In making these statements, I am not saying that farmers do not coöperate. Great numbers of them belong to organizations and societies of one kind and another, and the number is rapidly increasing. But the coöperation is usually not as complete as it might be, and very much of it does not originate from the land. Granting everything that is now done, there is still need of further effort.

(4) There is need of centers of interest in the localities, for lack of such interest is intensified by the rapid growth of cities and the directing of attention townward.

(5) Need of real personal starting-power and enthusiasm; of gumption; of enterprise that gets things done. Lack of this arises from little contact with fellows, from the arrested development due to marked individualism, and from the sterilization of rural institutions consequent on the removal of centers of interest to the towns. In the last analysis it is conditioned on the low earning-power of the average farm; but the earning-power is in the

hands of the farmer himself to a greater extent than are the social needs.

THE NATURE OF THE SOCIAL REMEDIES

As to remedies for the social shortcomings of the open country, only the most general suggestions can be given, but I think that it is fairly possible to indicate some useful points of view. Of course, the fundamental corrective of it all is education, but we should indicate what the nature of this education ought to be. We much need to know how to use our increasing technical knowledge, and to systematize it into practical ideals of personal living.

It is essential, as I have suggested, that we start with the proposition that farming people be kept on the farm. The centers of interest should be established or re-established in the open country itself, not further concentrated in the town or city. It is easy to see how interest converges in the city or town (page 16). Markets are there; roads lead there; trolleys and telephones lead there; the best churches, schools and entertainments are there. Farming is a local business: it rests on a particular piece of land; if the farmer is to be effective he must be content in his locality. The development of living local interest is the real root of the rural social question.

The importance of the personal and local initiative.

Every one of us, I am sure, feels that good institutions will not save us. Society can be saved and advanced only by increasing the number of competent persons who stand on their own feet. The farmer is proverbially the man who has stood on his own feet. Other persons have stood on other men's feet. The purpose of every good country-life institution is to develop persons who are able to walk alone. We must be careful that we do not develop a man who will go about his farming leaning with one arm on the government and with the other on the college or experiment station, and at every turn asking for recipes in franked packages. It is not the business of government to test every farmer's seeds, but to teach every farmer how to test his own seeds.

I think, therefore, that no agricultural work public or private, no institution state or national, no movement educational or philanthropic, has adequate justification unless its one purpose or effect is to allow native individual responsibility and initiative to develop in the man who stands directly on the land; and, if it is necessary to stimulate enterprise, the effort should lie preferably with the institution or agency that is nearest to the man and his problem.

Agencies of local communication.

The city has developed great effectiveness through its means of communication. The open country is just beginning to consider a similar phase of development. Undoubtedly, the engineer is to have a marked influence on the institutions of country life. Comfortable highways and electric lines are to open up the country. They will thread it with a network of avenues. Of themselves, these avenues will be concentrative or centripetal agencies, for the most part, piling up wealth in small cities and towns, as I have already indicated; we need to be thoughtful to develop at the same time the distributive or centrifugal agencies to counteract this tendency.

It is important that the lay of these avenues be such as to develop the country as well as the town. Good roads are a means of doing business expeditiously and economically; they are also a means of overcoming isolation, and they will have a great influence in organizing social movements in the open country. As all other avenues of commerce have been primarily city-feeders, it is of the utmost importance that country highways serve country necessities.

The distributive agencies will be largely social. The development of a country mail service is the beginning of an effective dispersive or disseminating institution. We must have a parcels post,—an institution that is opposed on the one side by express companies and often on the other by merchants in the smaller centers, who are afraid that persons will buy goods through the mails. It is humiliating that great public service should be obliged to wait on such opposition as this.

As nearly as I can estimate from such data as I have been able to collect, not one farmer in three reads an agricultural book, an agricultural bulletin, or an agricultural newspaper. It is all well enough that the farmer thinks in terms of experience rather than in terms of books; but a sound reading-habit is essential to his progress and his success. Reading-clubs, of one kind or another, are likely to become a strong force. It is not improbable that the agricultural press will find itself exercising a new coöperative relation with the reader. I look on the reading-center as one of the distributive agencies.

We should not forget that distributive agencies should be developed coördinately with the centralizing agencies, otherwise we make no permanent progress. It is not by any means sufficient that we have merely good roads and autocar routes and many trolley lines and a web of telephone communication. All developments depend for their final success on complementary movements.

Reconstructive movements.

Reconstructive agencies are already well under way. All the shift of which I have spoken is not without its decided reaction. The change of center has called for new kinds of institutions to stand for the country-life interests. These institutions are largely governmental. They reflect the rapidly growing tendency toward state or federal solidarity, and the delegation of power from the locality to the capitol. The rural population is now in danger of looking beyond its own institutions to government.

The greatest of all these new agricultural institutions in this country is the United States Department of Agriculture, working from the center outward, and gradually touching almost every isolated phase of country life. The growth of this great institution should be a source of pride to every American.

It now (1908) has disbursement of some $10,000,000; and every patriot, I hope, wants to see this sum greatly increased. Persons frequently remind us that this is a vast sum, forgetful or unmindful of the fact that it represents vast and fundamental interests. I prefer, rather, to think of it as a wholly inadequate sum, when I compare it with the $72,000,000 spent for the support of the army and the $102,000,000 expended by the naval service; for we must look for a time when departments that stand for peace by means of preparation for war will cease to exist, and when their regulatory and statecraft work will be distributed in those departments that rest on economic and social development.

The agricultural colleges and experiment stations present a decided recrystallization of agricultural ideals, constituting centers of influence more or less remote from localities, and representing a distinct centralization of power and of leadership within the states. They are rapidly becoming the centers of a new agricultural system.

The now common phrase "back to the country," has a deeper significance than most of us, I fancy, have caught. It means not only that certain persons are going, or talking about going, from the city to the country, but that initiative is being reflected from centers —largely governmental centers—back into the country. There is developing a ganglionic intellectual organization, as in years past there has developed a similar economic organization.

It is easy to trace the over-influence of the city in country affairs. The cities, for example, because they provide the market, are beginning to dictate the mode of producing milk, sometimes with too little consideration for the farming conditions. This work is properly a state function rather than a city function. The present agitation on bovine tuberculosis exhibits a similar influence.

The kinds of help.

I think that I see at least six classes of helpful activities for the betterment of rural conditions: (1) The discovery of local fact; (2)

the training of particular persons for special kinds of community work; (3) the organizing of the governmental function in agriculture; (4) the redirecting of rural institutions; (5) the developing of applicable education; (6) the appeal to personal leadership. We may now consider these classes sufficiently to enable us to catch their significance.

Aside from these efforts, we must remove all handicaps and disabilities that are not a natural part of the business, as the inequalities of transportation facilities, the effect of combinations in the interest of the few, discriminations in tariff and other legislation, the oppression of systems of marketing, the injustices of modes of taxation (p. 70). These and their kind constitute a very large subject, on the discussion of which it is not my purpose to enter.

1. THE DISCOVERY AND COLLATING OF LOCAL FACT

A thorough-going study of the exact agricultural status of every state should now be made, and it should be made by the state

F

itself, working through an agricultural college. Such an inquiry made carefully and without haste by men who are thoroughly well prepared, and continuing over a series of years, would give us the data for all future work with local problems. We must have the geographical facts. We are now lacking them. We talk largely at random. We must discover the factors that determine the production of crops and animals in the localities, and the conditions that underlie and control the farm life. Consideration of these conditions involves study of local climate; knowledge of the kinds, classification and distribution of the soils and the relation of place and altitude to production of crops and live-stock; determination of the best drainage practices on various soil types; consideration of the cultural experience and manurial needs as adapted to the types; inquiry into the practice with all leading crops and products of the localities; study of the possibilities for farm water-power; collation of community experience. Such a study of a state should be broad and general enough to consider the status of all the agricultural industries in the state, and it should also take full cognizance of educational and social conditions.

This constitutes the greatest need of practical farming at the present day. The agricultural institutions are working out the principles, but they may not be able to apply these principles to individual farms because they do not know the exact local conditions. The farmer himself may not know the principles, nor even the local facts. The result is a lack of articulation between the teaching and the practice. Farming is founded on the facts of the locality: no business can hope for the best success until it has exact knowledge of its underlying conditions.

Agricultural surveys.

These kinds of inquiries are now well under way in the form of "surveys" of many kinds, proceeding from the colleges of agriculture and the United States Department of Agriculture. The studies of larger range, that purpose to compare general agricultural conditions in the whole national domain and to standardize our knowledge of them, may well be undertaken directly by the national government; but the commonwealth itself should give itself the advantage of making inquiries into its own agricultural conditions. The survey work of the institutions will be greatly perfected in the next few years, and we may expect to see great public funds devoted to it. The survey parties will comprise strong, all-round men. No small part of the value of such surveys will be the discovery of great numbers of earnest, competent men and women on the farms who may be made local leaders, and the recognition that it will give to good agricultural practice everywhere. Every thorough survey should be the forerunner of new ideals for the communities, and of new points of crystallization of local effort. It should make new paths.

The model farm idea.

Many plans have been devised to develop this practical local experience. A notable suggestion has recently been made by high authority, advising the establishing of one model farm in every agricultural county, to be preferably under federal control. Passing the assumption that a 30-acre or 40-acre farm is the most desirable unit, and also the specific plan of rotations and style of farming, several underlying doubts may be expressed on the project: (1) Such farms would only indirectly utilize the natural and normal experience of the community: they would be essentially exotic or would at least be more or less arbitrary and imposed; (2) a "model" farm usually has little influence, since it is maintained under conditions that farmers cannot hope to secure; (3) the object-lesson method of teaching is not the most fruitful; it is not dynamic; it is proverbial that persons living near the very best farms, or about experiment station or college farms, may profit very little by them; (4) one farm in a county is by no means sufficient either to represent the dominating agricultural conditions of the county or to interest all the people in the county; (5) the farther removed the control of such a farm, the less can it hope to develop local experience wherewith to appeal to the people. The best model farms are actual farmers' farms. The experience on real farms, whether good or indifferent, should be assembled; this is a kind of advisory supervision that the state may very well undertake, working with the units and the conditions that are already in existence.

The model farm idea was dominant in the early days of the colleges of agriculture, but it has been found to be impracticable and one now seldom hears it mentioned.

2. DEVELOPING PARTICULAR PERSONS FOR COMMUNITY WORK

The failure of our fairest and most perfect plans traces itself to lack of good local leaders. In small towns and the open country there are club-houses vacant or of no account because there is no one person to organize and energize.

In cities, great things are accomplished by

settlements of one kind and another. Something of the kind can be done in the country, but it will need to be in the nature of better examples of actual farming as a base, with the farmer taking a new kind of enthusiastic interest in all the public and organized affairs of his community. The greatest aid will probably come by means of individual effort rather than by large settlement organization. The farming people must be reached through individualism rather than through institutionalism. Men and women may establish themselves as actual farmers, and while making a living from the land conduct a kind of social effort that is quite unknown in this country today. There is great opportunity for young persons to fit themselves for this kind of work, developing leadership and serving their fellows without the handicap of over-organization, which is likely to be a serious drawback in the highly specialized work of the cities. Nowhere will the individuality of personal leadership count for more than in the country.

It is important that the country work be

founded on occupation (that is, on agriculture), since all country interests rest on occupation. That is to say, the good social worker should be a farmer, rather than a missionary, charity organizer, officer of correction, or philanthropist. It is not a question of slumming. The rural people are not lost: they need opportunity and leadership. So far as possible, the work should be established in real rural regions, outside the towns. The worker should be resident the year round, not migratory; and above all he should not be of the summer boarder class. Inasmuch as personal leadership in country work must rest on a good foundation of agricultural knowledge, it follows that the best training place for this class of men is the agricultural school or college. Heretofore, these institutions have devoted their attention chiefly to technical agricultural instruction, but they are now rapidly taking up the social and the larger economic phases of country life. From some of the colleges, the young men and women go back to the country thoroughly alive to the necessity of organizing the social forces there. The time cannot be far distant when there will be some kind of a voluntary association between the students of all agricultural colleges, looking to the elevation of agricultural communities as well as to their own progress as farmers. Something like the student volunteer movement will eventually come out of this rising sentiment.

All public welfare societies should endeavor to interest good countrymen in the work that they are doing, bringing the countrymen into the organization, making them in effect local agents and representatives: this is very much better than to attempt to reach the problem by merely sending persons into the country.

3. THE GOVERNMENTAL FUNCTION IN AGRICULTURE

One by one the dominant affairs of the people are expressing themselves in administrational departments of government. Insurance, supervision of buildings, management of charities and corrections, engineering improvements, banking, railroads, and others, are represented by departments or bureaus having executive authority.

State departments of agriculture.

Every state in the Union, as well as the provinces of Canada, has some kind of a state-recognized organization, voluntary or otherwise, devoted to agriculture. Some of these organizations are societies; others are boards; apparently less than half of the organizations are really a department of the state government, and with perhaps a half dozen exceptions, these departments do not exercise extended governmental functions. For the most part, these state organizations are concerned chiefly with the exploitation of the agricultural resources of the commonwealth or with the holding of conventions and conducting of fairs. Sometimes, as in Michigan, Maryland and Colorado, the board of agriculture acts as a board of trustees for the agricultural college. In a number of the states, the composition of these boards is founded on representation by agricultural societies. In many of them, the executive officer is appointed by the governor. In a few cases, he is elected at the polls. These various modes and functions indicate that there is yet very little clear conception in this country of the governmental function of agriculture, as a definite part of a state cabinet; but this will come in time, as clearly as it has come in the administration of education, departments of health, and the like. Such departments will be frankly maintained by as large and free appropriations as those devoted to other parts of the state government, and the executive officer will be counted worthy as much salary as state architects, engineers and attorneys general. Such departments might have immense influence in dignifying country life affairs, in safeguarding them, and in stimulating the local initiative of which we have been speaking.

This means, of course, the new kind of public or governmental organization, one not conceived along political patronage lines. Government by influence must go, and government by merit must come. The conducting of such work must lie with men of ideals. It is time that specially trained men from the colleges of agriculture be put in charge of work in these state departments. If the agricultural college and experiment station idea is worth anything, it is high time that it be worked directly into state government.

The governmental function in agriculture is very much more than the making and executing of laws. The demand for fiats to exterminate bovine tuberculosis is a case in point. The real solution of this question is by means of popular education, conducted through colleges, schools, institutes and otherwise, combined with wise statutory regulation. In five years or less, any state could create public sentiment that would control the situation.

A new statesmanship.

Some of the largest questions now before the people are really rural questions; but many of the great rural questions—particularly those of a social kind—have not yet been recognized by public men. The solution of these questions will demand statesmanship of the very highest order. Statesmanship has been confined too much to so-called political questions. Many persons of first-class powers and thoroughly familiar with rural questions have had no opportunity to express themselves. I am convinced that the greatest present need in constructive statesmanship lies in the direction of agricultural affairs.

Attitude of state governments.

The lack of understanding of the relation of agricultural affairs to state government is well illustrated by the current attitude of such governments towards certain appropriations. Articles recently appearing in the press charge that directors of agricultural colleges and experiment stations are becoming politicians. If this charge is true it may be the direct result of methods of conducting state affairs. A state government is largely a business organization. It comprises departments that operate for the good of the whole. If a business man wishes to make his business thrive, he endeavors to determine what each department needs and appropriates to it all that he can spare and in proportion to its efficiency; and if any department is unable to do good work because of lack of facilities, he considers it good business policy to put that department in the way of accomplishing its best results. Now, the state often puts itself on the defensive against itself, as if under the necessity to repress its own departments. The result is that the director of an agricultural institution may feel obliged to organize his friends and become what is inappropriately called a "politician" in order that he may secure facilities to serve the state. It has been necessary for persons who have seen the need in advance, to do just this kind of pioneer work, but it would be unfortunate to oblige them to continue it. It would seem to be not beyond reason for the state to have an officer, commission or board,—as, in fact, some states have,—to make a yearly study of all state institutions thoroughly and to make recommendations as to comparative necessities, allowing the officers of such institutions to remain at home and develop their special work.

A state college or school of agriculture or experiment station is set by the state to accomplish certain work for the people; it is the duty of the persons in charge of such institution to acquaint the state government with the needs of the institution for the accomplishing of those ends; if it is not possible or expedient to supply such needs, the responsibility naturally lies with the government.

The state government part and the federal part.

I have said (page 75) that every governmental department or bureau devoted to agriculture should make its one purpose the developing of the personal initiative and the community feeling of the persons in the country; and that other things being the same, the department or group nearest home should have the greatest usefulness in this direction. As a concrete illustration of what I mean, I will cite certain types of work at present lying between the United States Department of Agriculture and the states.

Congress established a system of agricultural colleges in 1862. It established a system of experiment stations in 1887. It elevated the Department of Agriculture into an executive department of government in 1889. The colleges and experiment stations are state institutions, since the federal funds are given to the states. They constitute, however, the federal agricultural agents in the states. The federal organization is the Department of Agriculture. The concern of the colleges of agriculture has been largely the increasing of the productiveness of farming by teaching this phase of the subject and by inquiring for new truth. The duty of the experiment stations is necessarily to discover truth to the end that the land may yield more abundantly. The prime function of a national department of agriculture, as of other centralized government bureaus, is regulatory or supervisory. It deals with distinctly national questions, that is, with governmental questions. The United States Department of Agriculture, however, undertakes technical investigational work in the states, which may or may not be governmental, and for which the colleges and stations are now organized. There has been much demand for this kind of work and also much need for it, and the Department has tried, with great success, to meet the demand. The agricultural colleges and experiment stations once were weak because undeveloped. They are now beginning to grow and to come to their own. They are covering a broader field. Whatever it may have been necessary for the Department once to do, it may or may not be necessary for it now to do. In making these statements, I desire only to establish the fact that both opportunity and obligation lie with states and localities,—to urge not that the Department do less but that the states do more. It seems to me that we are under obligation to use our influence to relieve the Department of the necessity of doing some of the work that congressmen and others are disposed to ask of it.

What will be the ultimate relationship between appropriations for agricultural work by the Congress and by the states, I do not now propose to discuss. Federal appropriations will, of course, increase; and a way will be found whereby an increasing proportion of them may be so applied and disbursed as to stimulate local enterprise; and herein lies the possibility of the most fruitful species of federal and state coöperation. Congress might appropriate funds to be spent directly by local governments: the funds originate with the people in the localities. For the present, let us consider that there are regularly established agencies in all the states for the investigation of technical agricultural problems of those states. It is important that these agencies investigate these problems, not primarily because the agencies happen to be established, or because competent men happen to be connected with them, but because responsibility should lie at home with the people. No state can delegate to Congress the obligation of meeting its own problems; and every movement that tends to weaken local responsibility and initiative is a distinct menace to the people. Whenever the people are taught to look beyond their own institutions to federal institutions alone, they lose opportunity and power to help themselves. The people and the states are at fault in calling to Congress when they should call first to their own legislatures.

I conceive of only two usual reasons why the national Department of Agriculture should now be called on by the states to undertake technical agricultural work in the states: (1) When the institutions in the states cannot or will not undertake the work themselves; (2) when the problems seem to be regional phases of questions that have governmental bearings.

(1) In the first set of cases, it is equally the obligation of the state to handle its own special problems whether or not its institutions are able or willing to do so; but when the states are new or undeveloped, or when the neglect of the problems is likely to entail serious consequences on neighboring states or even on its own people, then it may be necessary to call on the federal government directly to aid or to interfere.

(2) Relatively few of the technical agricultural problems have true governmental or regulatory significance. The fact that the problems are common to many or even to all of the states does not place them in this category. Some technical problems have direct governmental significance because one state or even a group of states does not afford sufficient base on which they can be studied, and because they impose or necessitate regulation or elucidation by a central authority: the study of meteorological conditions is a typical example. Some problems are so expensive to investigate, that state governments may not be able to handle them. There may be certain other problems that need to be studied in all parts of the country in order that general and true conclusions may be drawn; these may be made the subjects of mutual and genuine coöperative study by persons in the localities organized to work consistently and harmoniously for a sufficient length of time to arrive at useful results.

Whenever regional information is desired by the government, on its part, there will be no difficulty in arranging for the securing of it through the regular federal agents in the states,—that is, through the agricultural colleges and experiment stations,—either from officers of the stations themselves or by federal officers delegated for the time being to the colleges or stations to work in the states. The college and experiment station will be glad to put their laboratories and facilities at the disposal of any investigator who wishes to come and make use of them. They would not think it right, however, to have independent laboratories or fields developed alongside, even though requested by persons in their own state or by the state department of agriculture,—not because of jealousy (for jealousy should be unknown to scientific men) but because such action would tend to diminish the confidence of its own people in the local institution, depriving the institution of the support it needs for the work for which it was created, and encouraging in the people a desire or willingness to shift responsibility. Much can be done and has been done wisely to strengthen the local institutions by the timely aid and suggestion of the national Department; but it is easy to see that such a policy might arise as a gradual result of the best-intentioned work as to prove in the end to be destructive rather than constructive.

States rights.

We have passed the old formula of "states rights." We have learned that certain things would better be delegated to federal agencies. Consolidation or centralization of power is a necessity. Yet at the same time we are pressed by the necessity of maintaining local initiative and vitality. It is possible to centralize power and at the same time to develop the locality— be it state, county, or neighborhood—if only we keep a clear distinction of functions. The real states' rights principle underlies the development of the individual and the community rather than the maintenance of the pride and prerogative of the commonwealth as an organism; that is, it means community privilege, duty and opportunity. It is properly a strong internal constructive policy. Such policies are more and more delegated to Congress, where fewer persons partake in them, and the states are not developing coördinately with the nation. Government should be kept at home.

We have talked much about states rights, but very little about state opportunity or coöperation between the states. We have tended to emphasize separateness rather than unity. The American, with his strong ideas of individualism, has really made less progress toward practical democracy in some directions than some of the monarchical peoples. The obligation of helping its people rests primarily on the state organization; but if the state will not render this aid, the federal government must do it. Of all affairs, the agricultural are the most native and local, and need to have the most careful concern of the state or community organization. The reasons for the recent growth of centralized federal power are at least of two classes: The growing urgency of inter-state problems; the failure of the states to meet all of their responsibilities.

The question, therefore, is broadly one of the governmental sphere and the local responsibility. We need to distinguish sharply between governmental function and investigational function, particularly when the people have already provided agencies for the two. I would not deny to any federal department either the right or the need to exercise the investigational function, but it seems to me that it is the secondary function and to be exercised on occasion. Any government department investigates primarily in order that it may coördinate, regulate, control, supervise, set in motion, and determine policies; it also needs to maintain research for the purpose of keeping its work vital and sound; but the great questions of organization and public policy are its particular sphere.

The coördinating agencies.

We may now enquire what are the proper agencies for the coördinating of the isolated and scattered forces of the open country, and for the more or less separate functions of state and federal governments. We first observe that these forces are of very many kinds. They constitute one public question only as they affect persons following a series of land-occupations; but the same forces may equally affect other persons. These agencies are of two great groups: Those that are educational; those that are regulatory or governmental. It is not necessary for purposes of administration that the assembling of all these rural agencies be centered in one bureau. On the other hand, there is distinct reason why they should not be so centered,—in the fact that agricultural forces are of right not isolated forces. Since many of them are broadly human, some of their significance is lost when we attempt to segregate them.

As real coöperative work crystallizes, the federal departments will have less need of maintaining independent relations with individual farmers in the states. They will deal with broader questions of policy and procedure as such problems are accumulated, the accumulation being very likely brought about under their influence or suggestion by the groups or agencies representing localities.

Coördination of educational matters.

Now, we can have no system and no sets of rules or methods to impose on posterity. There are too many schemes already. No man can determine the details for the future. All he can wisely do is to enunciate principles (if he has the penetration to discover them) and establish a few points of view. Many of our fairest schemes fail because of their very perfectness. It requires no foresight to say, however, that since what we are calling agricultural education is fundamental and not class education, and since there is a bureau of education of the national government, the coördinating of agricultural education should lie with that bureau. Agricultural education is in need of measuring and coördinating with education in general. If this function should not lie with such bureau, it will be because that bureau is incompetent to handle it, or is unsympathetic toward it, or is not given the necessary facilities.

In the states, the regular administrative departments of public instruction should handle the work of all fundamental elementary and secondary education. They will need to call on the agricultural colleges for help, especially in the training of teachers; but they should exercise the control. The trouble is, however, that such departments have not risen to this opportunity, and the agricultural colleges have been forced to take up the work, and the leading ones of these institutions are now doing all grades of educational service. Departments of education are likely to be followers of public opinion rather than makers of it. It is at this moment a serious question whether the regular administrative state departments or the colleges of agriculture are to carry the rural work; but the opportunity lies before the departments.

Education has now come to be a much broader conception than the work of formal schools. It covers a great range of activities looking to the training and developing of men. It is most unusual that in a country in which education is said to amount to a religion, there should be so little centralization of educational control and coördination at Washington. There is continual agitation for the establishment of new executive departments of government to represent special public interests. This agitation will likely increase. Much of the growth in governmental functions can be taken care of by enlarging the present departments, but there are certain great classes of progress and work that cannot be so accommodated; there are still great series of questions that lie outside the ordinary political region, of which public health is one. On the other hand, the President's cabinet is in danger of becoming too large. It occurs to me that there should be just one more department represented in the cabinet, and it should be of such nature that it can contain within itself all questions that will have to do with the general public welfare outside the field of regular governmental function as we understand it today; and this should be a Department of Education.

Coördination of agricultural matters.

The coördination of the real agricultural questions of national scope should lie, of course, with the United States Department of Agriculture. With the rapid growth in scope and influence of this Department and the vast agricultural interests of the nation before it, the place that this great executive organization is to occupy is beyond all conception. All our political and social future is conditioned on the resources of the soil. It is easy for the on-looker to see that the Department is gradually reshaping itself. Its work is becoming more educational, at the same time that it has acquired tremendous and permanent power in police work and in regulation. It is most interesting that the organization of this Department, with its sister, the Department of Commerce and Labor, should have been delayed so long. They represent finally the essential internal development of a nation; for the effectiveness of a nation rests on its farms, shops, mines, commerce, and the plain daily welfare of the people.

We must now consider what are the state agencies or instruments with which the United States Department of Agriculture is to coöperate in its coming organization work. Of course it should coöperate with all agencies; but there must be some, more than others, with which it can work most intimately and authoritatively. These agencies in the states are the colleges and experiment stations founded on federal grants; these institutions stand in much the same relation to the state, so far as leadership is concerned, as the United States Department stands to the nation. It might seem, at first thought, that the state departments of agriculture are the proper official channels through which general educational and organizational work should be accomplished. It is a fact, however, that these departments, speaking broadly, have not risen to leadership. Their proper field is closely governmental, inspectional and regulatory.

They do not have sufficient administrative freedom to undertake such extensional work as I have now in mind. Their organization and scope is quite unlike that of the national Department. They do not have the staff of trained scientific experts; nor is it likely that they will ever be provided with them, seeing that the leading states are now committing themselves unreservedly to the development of their agricultural colleges along these very lines. Moreover, with the necessary extension of governmental interference with agricultural affairs, these state departments will become more and more a part of the government of the states, and will have their attention occupied with very extended legal and supervisory functions; it will be a part of their sphere to coöperate with the national Department in these governmental functions.

4. THE RE-DIRECTING OF RURAL INSTITUTIONS

The great rural movement of the future is to be the evolving of a new social economy.

This is to be the lasting work of all national and state agricultural institutions. It is a work that is yet scarcely begun in this country. What progress has been made in this development has been mostly accidental.

The work of the agricultural institutions has been directed chiefly to increase the productiveness of the land—to make the farm earn more money. The agricultural colleges, for example, have properly laid their emphasis on this line of teaching; but in so doing they have themselves contributed to the maintenance of agricultural isolation. To make the farm more productive must continue to be the primary effort of these and similar institutions; but the time has now come when the colleges and all public agricultural agencies must join in the effort to improve and extend the social welfare of the persons who live on the land. The farmer is a member of the community.

In other words, while we need new knowledge, we need more than this to put the knowledge that we now possess into practicable and workable form; we must make it a very part of the men who till the land, and we must work out a means of working together. The greatest need is a radical revivifying and redirecting of all rural institutions. This is to take the form of a great constructive work, lifting the individual by developing the associative spirit in such a way that he may retain his own self-help at the same time that he secures the help of his fellow and the incentive of community action.

We must bear in mind the necessity for a change of face: we have maintained our position by means of vast extents of virgin land rather than by the excellence of our agricultural methods. It is interesting to note that one of the most matured European observers characterizes our farm management, particularly in the corn-belt, as "unparalleled in its wastefulness," setting up "a false economic standard in the industrial life of the agricultural classes, and will prove to be a bad preparation for the less bountiful times that must some day come." Every student of the economic condition must feel that the present unstudied or neglected type of individual farm management must be reduced to orderliness and effectiveness. The problem will arise here as it has arisen in Denmark, Ireland and other countries, unless we profit by their example and meet it in advance.

The necessity for working together.

A widespread system of coöperation must come for the open country. When I write the word coöperation, I use it in its true sense. I do not mean on the one hand a mere factitious business organization for buying and selling alone, or, on the other hand, so-called "coöperative work" of an educational or investigational nature with individuals here and there. Individuals pass; their influence mostly passes with them. Two individuals acting in unison or in coördination are more effective than the same two persons acting separately. If these two combine with another two, the effect is more than twice increased. True coöperative work with the Department of Agriculture or with an agricultural college operates with groups or societies of men, rather than with isolated men: the effect of the work is augmented, energy is conserved, and, most important of all, an organization is left behind to continue the work, or at least to advise the utilizing of the lessons that have been learned.

Any group-association that crystallizes about a real economic problem has the spur of necessity and therefore has vitality. All such local units should be known to somebody, and all of them should be organized into larger units. All of them should be encouraged, for they may be the germs of a new social order. Some central agency should coördinate and integrate them all so that, while every one maintains its complete autonomy, altogether they may progress toward definite social ends.

In New York, for example, there are more than two thousand creameries, skimming-stations and similar organizations, many of them more or less coöperative: What an opportunity to reach and energize the dairy industry! Some state agency should coördinate them on an educational basis; other states should coördinate their associations; the federal government should coördinate them all.

To assemble, direct, strengthen, to make effective the native coöperating expressions of the people, is an office the results of which are beyond all imagination. All agricultural experience, all experiments, all investigations made here and there by institutions established for the purpose, even all police work and the ordinary functions of government, should be assembled, solidified, and educationalized. It will be necessary for governments to send out regular agents or organizers for this class of work in the communities. In backward communities, it may be necessary for such agents even to organize coöperative creameries and other economic groups.

The economic organizations.

It is a poor country that does not support organizations to further its economic or business interests. We must have associations of some kind to further the interests of milk-producers, creamerymen, breeders, poultrymen, cotton-growers, florists, nurserymen, seedsmen, evaporated-apple men, fruit-growers, melon-growers, bee-keepers, horsemen, shippers, and the like. When such economic groups do not exist, it is the business of some one to see that they do exist; and, if they do exist, it is the business of some one to see that they are more effective. Society cannot escape the responsibility of being concerned in such group-associations. They register the effectiveness of the community.

Agricultural organizations have undergone an interesting evolution in this country. Some of the early groups were on the plan of a "society for the promotion of agricultural knowledge," apparently patterned after the so-called learned societies. The proceedings must have been ponderous. These appear to have been followed by the democratic discussion-society, still reigning in many forms. A social basis developed with the grange. Coöperative groups have arisen, but as yet with too little net results (p. 72); in some regards we are behind our European neighbors. The educationalized association is now developing, centering about the colleges of agriculture in the form of short-courses and "farmers' weeks." The continuing coöperating working society is yet to come with us.

The associations reflect the spirit of the persons who compose them. They ought to be the mainspring of all good works in the community. Every agricultural community ought to have something like a board of trade, holding regular meetings in a regular place. It is unfortunate that boards of trade are practically affairs of cities and villages.

What agricultural societies can do.

One of the commonest causes of discouragement in a farming business arises from the failure to utilize local or neighborhood experience. In every community there has necessarily developed a body of experience which should be of great value to the whole community if only it were collected and arranged so that conclusions could be drawn. Here is most useful work for any local society. In every neighborhood there is probably some person who has the ability and temper to enable him to collect and collate such experience. A great many of the questions that come to the agricultural conventions and to the agricultural colleges and experiment stations could be better answered at home if only the local experience were available for public use. In attending agricultural conventions, I am always impressed with the waste of effort in discussing questions of a purely local character. In considerable editorial experience, I have had the same feeling. It is necessary to answer over and over again the questions that have already been asked and answered over and over again, until one almost comes to feel that the work is not progressive. I must not be understood as advising that these questions be not asked or answered: I only wish that we had the means of answering them more definitely and in the place where the answers would have significance. There are different kinds of questions to be asked in different kinds of conventions.

Every agricultural society needs to emphasize the public-service phase of its work. This phase of society work is not yet understood in this country. Fifty years from now it will be the dominant note in all rural societies. I am quite sure that I have not the power to make my meaning wholly clear or to convey a vivid picture of what I have in prospect. We are so unaccustomed to thinking of such subjects that we have not yet developed a point of view or a vocabulary.

Consider, in the first place, that practically every man has stood alone in his farming, and has been obliged to contend with all the organized interests of the business world. The result is that he is a negligible factor in trade, in a great part. What is true in business relations is more broadly true in social relations. Our present greatest need is the development of what may be called "the community sense"—the idea of the community, as a whole, working together toward one result. We must admit that there is now a deplorable lack of any associative effort that commands respect and puts things through. This community sense must be accomplished, as I have suggested, by the organizing of many local societies or clubs, and the coördinating of these into larger societies. If the individual farmer, working alone, is a weak economic and social unit, so the isolated society or club is also weak.

It is not my purpose to present any plan of developing the community sense, but only to enforce its necessity to a better and more fruitful country life. As a practical matter, the developing of an effective community feeling must rest with the leadership of some one strong organization. I once suggested to a noted horticultural society that it might well determine for its members many of the vexed questions concerned with the varieties of fruits by establishing demonstration or volunteer orchards. Then I suggested that it had a privilege and a duty touching useful horticultural education of a collegiate grade. Later, I suggested that such society should have a continuous working existence throughout the year, engaging in the organizing of subordinate or contributory groups. It is a question whether, in these new days, a yearly convention is sufficiently effective. The time has come when we ought to distinguish between conventions and organizations. The reason why the grange is so very strong in the states in which it is well organized, is because it is in operation continuously. Association is assuredly the keynote of the future; but the association of which I speak is very different from the labor-union kind.

The coöperative spirit is far broader than that expressed in formal organizations. It is also the basis of good government.

Possible extent of associative work.

Every kind of organization that now exists in the open country, and which can be readily extended to the open country, may be made the means of carrying the gospel of coöperation, companionship and better farm life to the persons who live on the land. A new meaning must be given to societies. No society should be maintained merely for the purpose of entertainment, but it should have vital relation to the real affairs of the community of which it is a part. The number of rural organizations and associations is surprisingly large, even not counting the technical agricultural societies and groups (which are really the most effective of all). It is not so necessary to organize new groups as it is to fertilize and redirect the old ones. Rural institutions ought to be effective because they are, for the most part, natural expressions of indigenous needs, the outcome of the community's work. Many of the city institutions are creations of some person's philosophy or the expression of some fad or fashion, and they are likely to be imposed on the community rather than to grow out of the community.

Let me enumerate some of the group-associations that might easily aid in the regeneration of country life if they were not so closely tied to their customs and traditions: The school; the church; fraternal societies of all kinds; christian associations for men and women and for both; all singing schools and musical clubs; reading clubs and library associations; women's clubs; historical societies; athletic organizations, and all groups that might develop the play spirit and revive the native games; local political organizations, that might give as much attention to developing or promoting the community as to putting some one in office; the good roads interest, which is now easy of organization and direction; banks, which might have relation to the welfare of the community as well as to themselves, as is shown by various European experience; chambers of commerce and business men's organizations in the smaller cities, that might extend their efforts beyond the corporation lines, associating the country merchants and traders with them; civic societies; improvement and art societies. In the way of economic group-associations, the coöperative creamery may be cited as a representative example. In a dairy country, such an institution not only works an improvement in farm practice and develops a market, but it makes a meeting-place and thereby should come to be a useful center of local interest.

Rural government.

When, from long association, we become accustomed to an institution, we lose our habit of challenging it. It is a question, for example, whether we do not need a redirection in rural government. The rural people are not inert, as they are often said to be, nor are they incompetent, but the systems whereby men are organized and affairs are directed are likely to be incomplete, ineffective, and to lack vitality. I think we need more active and compact rural government. I am afraid that some of our systems of governing the open country may be found to be antiquated and inadequate.

Community government should not devote its chief attention to mere regulation and to support of defectives and dependents. Its first concern should be to set productive forces in operation. One community should emulate another. Even state legislatures are likely relatively to disregard constructive enterprises.

Application of public money.

Again, the agencies that will develop the institutions of the open country will rest on a type of mind of town folk quite as much as on the activity of the country folk. The town folk should consider it to their interest that the drainage cityward does not go too far. The town must recognize its obligation to the agricultural country. In New York, for example, 85 per cent of the taxes are paid by Greater New York and Buffalo, notwithstanding that there are probably a million farm people in the state, and a farm property valuation of much more than one billion of dollars. But this great fact constitutes no reason of itself why these cities should control the distribution of the tax money of the state, or even of all the tax money that they themselves contribute; for the wealth of the cities did not originate in the cities. A good part of it has come from farms in New York state and elsewhere. The farmers have given the city traders an opportunity to trade, and often to make more money in the mere trading than the farmer made in the producing. If there have been in-gathering agencies, so also should there be distributing agencies back to the sources.

It is a wrong philosophy that would apply the proceeds of taxation only to the localities in which they originate. The state is an organism, and cities, like the country, are only parts thereof. Whatever may be said for or against strong centralization of government, it has the tremendous advantage of being able to expend the revenues collected of all the people in the interest of all the people. All along, the cities seem to have carried the idea that the country is answerable chiefly to them; but the city, also, is equally obligated to aid its contributory country—to do its share in the furnishing of public revenues wherewith to build country highways, country churches, country schools, and other rural institutions, and at the same time to allow the country the controlling voice in the disposition of these funds.

Something of the same kind may come, as I have already suggested (page 98), in the apportioning of funds by Congress back to the people with whom they originated, giving to the people in the localities the responsibility of using them. The Land-Grant Act of 1862, establishing the colleges of agriculture and mechanic arts, is of this nature; so is the Experiment Station Act of 1887, although the state autonomy is less complete in this case.

Banks.

We need carefully to consider the migration of money. I have already indicated that money tends always to pass out and away from the place of its origin. It is multiplied as it goes; but a good part of it ought nevertheless to describe an orbit and come back to its locality. Better still, a good part of it ought to be kept in the locality for local uses. Banking institutions ought to be agencies for more directly developing the region of which they are, or ought to be, a part, by keeping sufficient money moving in the region to supply its needs freely. These institutions now exist, in this country, for the stockholders, who are usually not producers. I am always impressed with the palatial buildings that banks are able to erect. In other countries there are coöperative associations that find money for their members to use in the making of the crops. Because of the prosperity and consequent independence of the American farmer, the lack of loanable capital has not been a serious handicap, except perhaps in the South, but he could nevertheless use such capital to very great advantage.

A fault with our banks, considered from the standpoint of the development of the community, is the fact that they loan only on property, thereby eliminating the poor farmer just in the proportion of his needs. Moreover, they loan on too short time, as a rule, to cover the making of a crop. The result is that the farmer is driven to the merchant and the usurer. In the South, where the lien system has been in operation, the merchant is likely to refuse to loan to a man who desires to change his system of farming, even to improve it, fearing that it may be only an experiment; the result is that the lien system becomes a preventive of progress.

Some of the European systems have demonstrated that it is perfectly feasible and safe to loan money on the industry and honesty of members, thereby not only furnishing the necessary funds but putting a premium on character and thrift, and keeping the money at work where it is needed. The coöperative systems find it advantageous to employ agents to instruct their members in farming, becoming thereby a social and educational force in the locality. Our banking systems are devised for the handlers of money, whereas some of them, at least, should be devised for the workers and common users. Banks, as well as government and schools, should be native.

Fairs.

The fairs are one of the anomalies of the present time. If one is interested in the evolution of institutions and the persistence of customs, he can find here material for long and rewarding study. In great part the fairs are meaningless and are not agricultural institutions. They display the unusual and abnormal, they exhibit the prodigies, they encourage a class of professional exhibitors, they attract the gaudy and the doubtful; they are, in fact, largely exclamational. There is permanent discussion as to the allowing of racing and betting and of mere performing; but these things are really symptoms after all, and the real solution is to redirect the whole enterprise.

County and local fairs might well be part of a thoroughly organized state system, and be taken out of the influence of showmen and race-track gamblers. The fair should be a kind of school, and its work and influence should exist continuously throughout the year. The fair ground itself should not be idle fifty-one weeks every year. Its facilities could be used for many exhibits and schools at other times; and a good part of the grounds could be utilized by school children or others to grow plants that should stand in exhibition and teach a lesson when the fair comes round. It is not too much to hope that fair grounds may some day contain school-gardens. Such an associative organization should go with a fair as will keep intending participants—I do not like the word exhibitors—in a state of preparation and attention for the whole year. The fairs should reach all farm children. And the significance of everything that is shown at a fair should be explained by a good teacher standing on the spot.

The rural church.

I have said (page 17) that the country church is not accomplishing what it might for the rural communities. This is not due to lack of devoted service on the part of the country pastor, but to a need of re-direction in the institution. Concerned in too many cases with technical religion, formal piety, small and empty social duties, the country church may not appeal strongly to men with rich red blood in their veins. Country people are hungering, in many regions, for wholesome spiritual guidance, although they may not know it. They can be reached by an energetic church. The church must do many kinds of extension work. It should express and encourage the natural inspiration that may be made to flow from the common affairs and practices of any agricultural community. Are not the plants growing and the animals thriving? Are not the crops going in and the harvests gathering? Are not the winds blowing and the rain falling? Are not these real things, that every person understands and that can be made the means of reaching every man who has them? Yet all these things are practically a sealed book to the church. The rural church buildings are essentially what they were fifty years ago,—a preaching room and a vestibule. Why not make a country church a social center, letting it stand for good works in everything that interests the community, and placing it in some direct relation to vocation?

The church is the oldest and completest of organizations. It is deepest in the affections of the greatest number of persons. We cannot afford the waste of effort that results from the decadence of the rural units. There is urgent need that we utilize this institution for the quickening of country life, making such life worth the attention of educated young men, and developing its natural and legitimate social attractiveness.

There is no greater field for service than in the country church. Young men should be prepared consciously for this service. A good part of the training should be in social questions in their rural phases. A course in a good agricultural college might supplement the training in the theological seminary. Religion should be native. It should be concrete and applicable. Religion is the natural expression of living, not a set of actions or of habits, or a posture of mind added to the daily life. The type of religion, therefore, is conditioned on the kind of living; and the kind of living is conditioned, in its turn, very largely on the physical and economic effectiveness of life. The religion of the open country should run deep into the indigenous affairs of the open country. Everything with which men have to do needs to be spiritualized. This is much more effective for our civilization than merely to spiritualize things that we hope for.

5. THE DEVELOPING OF APPLICABLE EDUCATION

It is not enough that education facilities be provided for all the people: the education must have meaning. It is true, of course, that the function of education is to develop and train the mind; but the mind may be trained by means of many subjects, and some subjects or processes are best for one group of persons and other processes for other groups. As farming is a local business, so is it specially important that the affairs and objects of the locality be made part of the means of training the farmer.

Education is not confined to the institutions known as schools. It is the result of all experience and all training. Many new agencies are contributing directly to this training or are modifying its application. Some day the school will utilize and direct the experience that the child gains in its home life, as well as in its school life, toward a distinct educational end.

Special adaptation of the school to the needs of farm laborers is now very much needed. Neither the agricultural college nor the ordinary type of rural school can reach this end. Farms are not training farm artisans as the shops are training shop artisans. The shops are amongst our very best schools. The farm-artisan school must be taken into the locality where the artisans are. It is doubtful whether it can be a night school. It may have to be a local winter school taught in a new way as an adult school, until the farms themselves are good enough and well enough organized to be schools to train their own men. The reading-clubs now proceeding from a few colleges of agriculture suggest the beginnings of a new movement for community and home education of this kind.

Necessity of new point of view.

Although we are properly proud of our public schools, we should not cease to challenge them. We are now in the epoch of the domination of the schools by the colleges. Narrow literary college entrance requirements are wholly out of character with the present necessities of living. As soon as the public schools begin to connect with the industries, all antiquated kinds of tests must go; I look for secondary teaching in agriculture to help in this, although at present we have no schools to articulate with the people on the one hand and the colleges of agriculture on the other. The high school has been over-developed in its prevailing form. It probably represents the end of a line of social evolution, and we may need to go back and take a new start.

The schools are dominated too much by system and regularity. The control by state departments is constantly becoming more rigid. Our effort seems to be to make the educational processes uniform for all pupils, notwithstanding the fact that no two pupils are or ought to be alike. There is now agitation for more thorough supervision of the rural schools, and such supervision is undoubtedly necessary; but if this supervision is to result in complete domination by a central authority at the capitol, we can well afford to wait.

There seems to be little personal life-motive in our education. The process produces passive or static results. It does not seem to develop the quality of leadership, as it should. We over-emphasize the importance of mere verbal accuracy and breed in our pupils a depressing fear of making mistakes. The schools do not send their graduates home to work in village-improvement societies, civic clubs, farmers' organizations, mosquito extermination, or to give them a just point of view on the common affairs of the community. Part of this lack is no doubt due to ineffective home training. With the growth of leisure in cities and towns, children are not trained to the responsibility of work. They read novels and entertainment-periodicals, and are afraid of soiling their clothes. We are developing a kind of artistic idleness. The pity is that it is considered to be respectable.

We must outgrow the sit-still and keep-still method of school work. I want to see our country school-houses without screwed-down seats, and to see the children put to work with tools and soils and plants and problems. A child does not learn much because he is silent and inactive. Out of this work will grow the necessity of learning to read and figure and draw.

This redirection of educational effort will, of course, come slowly. A new spirit is arising among school patrons here and there, and this will aid to bring it about. The people must feel that they, themselves, have something to say about the schools, and that all power is not centered in departments and boards. And then society should see that useful ideas of education are sown among its members.

If these redirective forces are to be set in motion and made effective, much public money will be required. This money will not be a gift to an agricultural class, but an appropriation to aid in developing the internal resources of the country. The farmer is not a subject for charity; but it is due him that handicaps be removed and that he be allowed full opportunity for development. Special legislation of many kinds, from tariffs to corporation laws, have favored other and centralized interests, and have encouraged the growth in many quarters of colossal selfishness. The farmer has been the forgotten man. What we appropriate to him for education and knowledge is a small offset to the special privileges that have been given to other men; it is his peculiar recognition from government, and education, in the end, should produce a more wholesome result than any other aid that government can render to its citizens, and this should inspire the best kind of voluntary effort.

Importance of the rural school.

In our eagerness to serve the agricultural interests, we are likely to place relatively too much emphasis on the importance of establishing of new institutions, whereas the greatest effectiveness and even the quickest results may very likely be attained by utilizing the agencies already in existence. It is easy, for example, to ridicule the country school and then to plead for new isolated schools in which to teach agriculture; but in so doing we may forget that isolated special schools cannot serve all the people and that they also tend to isolate the subject. The present rural schools, with all their shortcomings, are good schools because (1) they are already in existence, (2) they are the schools of all the people, (3) they are small and thereby likely to be native and simple, (4) they are many and therefore close to the actual conditions of the people. I would utilize them to the fullest; and in the end these schools, when redirected, will present the solution of the problem of rural education. In the remarks that follow, I mean no criticism of teachers in the rural districts. From long association with them, I have come to regard them as a devoted class, and they comprise some of the best teachers that I have known. They deserve every recognition and encouragement.

A consideration of the school question will enable me at once to illustrate what I mean by the redirecting of rural institutions and also allow me to suggest the relation of such redirection to local pride and initiative. These rural schools fail because they do not meet the living needs of the people. They do not teach the objects and affairs of their environment. They are not vital. But in all this they differ from all other schools only in the fact that their progress is somewhat slower. Neither are city schools often really vital. Neither, perhaps, is the greater part of our college instruction. Until very recent years even the agricultural colleges have not taught vitally. The public schools do not yet teach the essentials. The first object of any school should be to teach persons how to live.

But with all their shortcomings, the rural schools are really making progress; and I am sure that some of the speakers whom I have recently heard do not know how considerable this progress is. Unfortunately, it cannot be recorded in statistics; it is a new atmosphere, a progress in the spirit of the school and in a somewhat changed outlook rather than in the adding of subjects under new names.

Three movements looking in some measure to the supplanting or reorganizing of rural schools are now well set in,—(1) the granting of aid by Congress for the schools of the people, (2) the consolidating of existing schools in order to increase the efficiency of teaching, (3) the establishing of secondary schools for agriculture. All these movements will contribute to the solution of the rural school problem, but it may be well to examine them somewhat carefully.

(1) *The subject of federal aid.*

Bills are before Congress to grant federal aid to secondary and normal schools in the states. It is a new policy. Its consequences should be carefully considered.

It is relatively easy to secure money from Congress, because it is least accountable to local public sentiment and because its funds are not derived from direct taxation. The tendency is to go more and more to Congress for funds, even for purposes that should be cared for by the states and the localities. We must be careful to teach that persons do not rely on some one else or on government. Congress may well undertake new work in the states for the express purpose of showing the way and stimulating local ambition when the work is of such magnitude that states or localities cannot undertake it. The establishing of a system of agricultural colleges and experiment stations is a case in point. It is quite another matter, however, for work that originates in localities and belongs to them to go past the commonwealth and to appeal to Congress. If states and provinces should exist at all, they should take care, as much as possible, of their own internal development.

It is indisputable that the common schools need more funds. Part of the funds should come from the localities themselves. Part should come from the state. Whether any part should come from the federal government, thereby expressing the national spirit in education, is perhaps largely an academic question; but the methods of its disbursement and control, if federal money should be given, is a subject of the first importance to the maintenance of local institutions. The opportunity of full initiative and independence should be popular rather than official; and every safeguard should be taken to see that national schools are not forced into close uniformity.

The lack of pride and gumption in our rural schools is probably already due in no small part to the removal of motive-power from the locality to the state capital. I have in mind a rural school on the premises of which a neighbor had placed personal property. For years the teacher and members of the local board and isolated friends had complained and threatened. The material remained. Finally a visitor reported the matter to the state authorities some two hundred miles away, and the nuisance was forthwith removed. A school locality that is impotent to remove a pile of lumber is also impotent to make very much progress in its schooling.

The small effect of "arbor day" is also most surprising. After all these years of planting, and of song and recitation about it, the communities have not yet risen to the point of having well-planted school-premises. The larger part of the grounds are yet bare of good trees. This would not be so if there were any genuine local interest in the subject of improvement of school-grounds.

Whether the federal government may properly aid the common schools in the several states is much more than a question of state or community opportunity, however. It is often a question of community ability. A child ought not to be disadvantaged by the locality in which he lives. All should have equal educational opportunity. The ratio of population to taxable property and to the labor demand, differs widely in different communities; and society must in some way see that chances are evened up in the communities that cannot support proper schools. How far this shall be done by state and federal agencies may perhaps be a matter of detail; but it should still be possible to develop enterprise and responsibility at home.

(2) *The consolidating of schools.*

The arguments in favor of consolidation of schools are many and important. By consolidation, stronger teaching units are secured; more money is available for the employing of teachers and the providing of equipment; special subjects can be given adequate attention. The objections are many, and most of those commonly urged are trivial and temporary. The greatest difficulty in bringing about the consolidation of schools is a deep-seated prejudice against giving up the old school. This prejudice is usually not expressed in words. Often it is really unconscious to the person himself. Yet I wonder whether right here does not lie a fundamental and valid reason against the uniform consolidation of rural schools,—a feeling that when the school leaves the locality something vital has gone out of the neighborhood. Local pride has been offended. Initiative has been removed one step farther away. The locality has lost something. It is a question, even, whether the annual school meeting is to be lightly surrendered, whether it is not worth keeping as an arena for the clearing of local differences, and as a possible nucleus of a useful institution.

I would establish an institution in every locality rather than banish one from it. I should like to see on every important four corners in the open country four institutions,—an assembling place, as a town meeting hall or grange hall; a building that stands for the products and history of the community, into which could be gathered a local museum, historical mementos, biographies of the inhabitants, and in which there might be a useful library; on another corner a redirected rural school; and on the other a redirected church that should strike its roots deep into the native affairs of the community.

I fear that much of the impulse for the consolidation of schools is a reflection of the centralized formal city graded school; but it is by no means certain that these institutions are to be the most important or dominating public schools of the future. The small rural school, with all its weaknesses, has the tremendous advantage of directness and simplicity. It is a great question whether it would be improved by a rigid system of grading. It is a question, in fact, whether the present graded schools do not still carry the onus of proving themselves.

Unquestionably, consolidation of rural schools is often advantageous. It is to be advised whenever it seems to be necessary for pedagogical reasons. It is often urged, however, for financial reasons; but this in the long run, is not reason enough. We maintain our canals and government work at public expense. The state must coöperate in the maintainance of its detached schools, by direct appropriations, if necessary, to their localities, always on the condition, however, that all effective control does not pass out of the community. Consolidation of schools is much more than an educational question. It touches the very quick of local pride and progress.

(3) *Special agricultural schools.*

I speak now of the separating of education in agriculture. My readers know that many years ago there was long-continued agitation for agricultural and other industrial education. Necessarily the discussion took issue with the existing order of education. The movement was essentially a revolt. This long-fought revolution culminated in the Land-Grant Act of 1862. The collegiate grade or phase of agricultural education was established forever. This new education was so unlike the old education in spirit that new colleges were established independently of the old. The new education was isolated. In some instances, the new education was made a department in old institutions, but in such conditions it did not thrive. The separate colleges led the way. Being free, they could do as they chose. They did not need to conform to old customs and methods; yet it is worthy of comment that, although being free they were nevertheless bound, for they carried the new work as a recitation-subject and book-subject and lecture-subject, following the traditional pedagogical systems. Long the new education lay in a pupa stage. The mechanic arts phase first got its wings. Now the agricultural phase is bursting its shell. But we find that the separate colleges no longer hold the exclusive leadership. We have found that education that makes use of agricultural subjects is education, just as much as that which uses mathematics or mental philosophy; and this being the case, they all live together in harmony. "Culture" and "mental discipline" are not mere abstractions: they are the results of good concrete work.

Not only do they live together in harmony, but all of them gain much from the association. This new education has even put a new attitude into much of the literary teaching. Moreover, I am very sure that the new industrial and personal education has saved the old college and university from extinction; or, to put it in another way, that if they had not taken it in, the evolution of our present-day collegiate education would have arisen on new foundations, and in time the old foundations would have been left hopelessly stranded. Perhaps some of the old-fashioned institutions that are isolated in spirit are stranded now; but they may not know it.

In the meantime, the separate agricultural colleges have maintained themselves, but they are no longer separate in spirit. They have allied themselves so far as they are able, with all public movements looking to the betterment of the "industrial classes," as the Land-Grant Act states it "in the several pursuits and professions in life." The separate colleges have their work to do as heretofore, and they will increase in efficiency, but they are special institutions, standing apart. We now see that these colleges, good as they are, do not satisfy the needs for collegiate training in agriculture, although contributing towards that end; for all the people must be served, agricultural education is, properly, not class education, and all institutions, on their own account, need the human and contemporaneous spirit. In all institutions of the people there should be opportunity for training in the affairs of life. The agricultural department or college of an existing so-called "liberal culture" university or college has been able, so far as it has been efficient, to drive home the personal and vital subjects and to cause them to be recognized as a coördinate part of a broad education. The rise of public sentiment, and the growth of wisdom among educators, are welding the old and the new: the agricultural teaching is being liberalized; the traditional teaching is being practicalized.

Just now another movement for personal education is well set in. It is the movement for the teaching of agriculture in the common schools. Significantly enough, it is mostly a separatist movement. We are attempting to isolate it by establishing separate agricultural schools or by organizing separate classes in existing schools. The establishing of separate schools is repeating for the common schools what has been the history of the development of the colleges. It assumes that the existing schools should not teach agriculture or that they cannot teach it effectively, and that such teaching should be isolated and that it exists for a class. Persons in the localities in which the separate schools are established will benefit by them. Persons who attend other schools will be debarred the privilege of being taught in terms of the country environment.

I hold that education in terms of the environment is the right of every citizen; and in the open country this kind of education is agricultural education, whether so called or not. But every citizen can secure this privilege only when he can have access to it in any of the public schools; we know that all the public schools together can barely reach all the people. In New York state, for example, there are some fifty-five agricultural counties. There are some 227,000 farms. If each of these counties had an agricultural high school graduating fifty pupils each year, to give only one boy from each farm in the state an agricultural course would require eighty-two years; and new generations are coming on in the meantime.

It is well to consider what the effect of this system of isolation will be on educational policy. The people will patronize these agricultural schools because they will be useful and significant schools. More than most other schools, they will teach the essentials,—that is, they will teach persons how to live. More of these schools will be demanded. A duplicate system of public education will arise. It is easy to see the ultimate result: if the common schools do not redirect themselves, they are lost.

I mean to say that the common schools need agriculture in order to save themselves. Of course, I mean agricultural education in its broadest and rightful sense,—the training of a man by means of country life or rural subjects, not merely the making of farmers. My old-time school friend will laugh at me when I tell him that his school is in danger, but I cannot be mistaken, and for the very good reason that his school is inadequate.

Gradually, however, we shall find the public schools readjusting themselves. They will reach out and take in the essentials. Then training by means of agriculture will take its rightful place as a part of a normal and natural school system, by which all the people everywhere—and not alone in some isolated school—may benefit. Education by means of agriculture is public education.

And the separate agricultural schools? Well, we may prophesy from the experience of the separate agricultural colleges. These separate schools will be wonderfully effective, so far as they are in the hands of men who are not tied to old points of view. For some years they will hold the leadership. They will develop public sentiment. They should always remain most effective agents for certain kinds of teaching. But they will always be more or less special schools. They will not satisfy all the needs for school training by means of agriculture. There will be some states and localities that will not establish them, and these localities will be considered to be behind the times. Moreover, such isolated schools are likely in the long run to deaden initiative in the many other localities that may most need them; and they are one step further removed from the people. So far as they tend to vitalize, by their example, the whole native school system, in so far will their effectiveness be beyond dispute; and this of itself will be worth all they cost. They will be pioneers. The real and lasting progress, however, is to be made by those localities that first completely redirect the existing schools in the interest of all the people.

The redirecting of the rural school.

Having now considered some of the new external influences that are likely to modify the common schools, I may explain what I mean by the redirection of the schools themselves. All effective education should (1) develop out of experience; (2) this experience should have relation to vocation or to the pupil's part in life; and (3) every school should be the natural expression of its community. If these statements are accepted, then it will be seen that the mere addition of a subject here and there to the school curriculum may not be sufficient to put the school into real relationship with its environment.

I am in sympathy, as I have said, with the establishment of a few secondary special schools for teaching agriculture whenever they can be well organized and the subjects thoroughly well taught. I am also in sympathy with the introducing of agriculture as a special subject in rural schools whenever it can be effectively handled. These two agencies ought to be effective in arousing and crystallizing public sentiment to the need of a new kind of education. However, these cannot meet the problem of rural education. The final ineffectiveness of merely adding agriculture to the curriculum lies in the fact that it does not constitute of itself a real redirection of the whole point of view of the school, although it may be a most useful means of starting a revolution that will bring about that desirable end.

If we establish special schools for agriculture, they should supplement the public school system, affording opportunity for somewhat advanced training to those pupils that are particularly interested in the subject; they should never be the instruments of diverting public attention from the necessity of allowing every school and every pupil the advantage of training by means of agriculture and industrial subjects.

I am afraid that we are accepting, without question, the present method of the high school and college, particularly its laboratory method. In the argument for separated rural schools I am always struck with the plea that good laboratories may be secured. A good part of this argument comes from college men. It does not at all follow that our four-wall laboratory methods are as useful for the secondary schools as for colleges and high schools. In fact, it is a question whether much of our laboratory work is really worth the while, as compared with good natural field-work under the conditions that are everywhere at hand. The agricultural schools and colleges, of all others, should develop the highest kind of nature-teaching. Yet they are likely to follow the tendency of the time and to produce a class of teacher that is dominated by the formal laboratory. I cannot help feeling that the greatest professors of agriculture or agronomy or horticulture or animal husbandry will be of the field-naturalist type. Laboratory-teaching may be pedagogically just as incorrect as book-teaching.

It is not necessary to have an entirely new curriculum in order to redirect the rural school. If geography is taught, let it be taught in the terms of the environment. Geography deals with the surface of the earth. It may well concern itself with the school-grounds, the highways, the fields and what grow in them, the forests, hills and streams, the hamlet, the people and their affairs. We are now interesting the child in the earth on which he stands, and, as his mind grows, we take him out to the larger view. A good part of geography in a rural community is, or should be, agriculture, whether so called or not.

Similar remarks may be made of arithmetic. The principles of number are everywhere the same; but there is no reason why practice problems should not have local application. In my day, at least, a good part of the practice problems were mere numerical puzzles. I fancy that even at the present time the old people are interested in the problems that the child takes home merely because the child is in a fix and his predicament appeals to their sympathies. When, however, the child takes home a problem that has application to the daily life, there is a different attitude on the part of the parents, not only to the problem, but to the school that gave the problem. A good part of agricultural practice can be expressed in mathematical form. How to measure land, how to figure the cost of operation, how to compound a ration or a spray mixture, how much it costs to fight bugs in the potato field, the mathematics of rainfall and utilization of water by plants — these, and a thousand other problems that are personal and sensible, could be made the means of so redirecting number work as to make it a mighty force in putting the school into relation with the community.

My reader can at once make applications of this line of thought to the reading, to the manual training, and to the other customary work of the school. Manual training develops chiefly skill: it does not articulate with life. The study of history should result in better local civic ideas. Text books err in merely making applications of their subject here and there: they need a complete change in point of view and in method.

You have only to consider the school-houses to see that the rural school is in a state of arrested development. Go with me from Maine to Minnesota and back again and you will see in the open country practically one kind of school-house, and this is the kind in which our fathers went to school. There is nothing about it to suggest the activities of the community or to be attractive to children. Standing in an agricultural country, it is scant of land and bare of trees. I think that if a room or wing were added to every rural school-house to which children could take their collections or in which they could do work with their hands, it would start a revolution in the ideals of country-school teaching even with our present school teachers. Such a room would challenge every person in the community. They would want to know what relation hand-training and nature-study and similar activities bear to teaching. Such a room would ask a hundred questions every day. The teacher could not refuse to try to answer them.

The problem of the rural school is not so much one of subjects as of methods of teaching. The best part of any school is its spirit: I can conceive of a school in which no agriculture is taught as a separate, which will still present the subject vitally from day to day by means of the customary studies and exercises. The agricultural colleges, for example, have all along made the mistake of trying to make farmers of their students by compelling them to take certain "practical" courses, forgetting that the spirit and method of the institution is what sends the youth back to the land.

Of course this new school method will demand trained teachers, but it should be no more difficult to train them into this point of view than as mere specialists.* The whole enterprise needs to be developed natively and from a new point of view; for in an agricultural country agriculture should be as much a part of the school as oxygen is a part of the air. I would not isolate agriculture from the environment of life in order to teach it: I would teach the entire environment.

The agricultural colleges and experiment stations.

Ten, or even five, years ago I might have said that there were need of redirection in the agricultural work of the colleges of agriculture and mechanic arts, but happily these institutions have now slipped their academic bonds. They are getting hold of the real objects and the real

* My own view of the ways to secure teachers for these subjects is expressed in Bulletin No. 1, 1908, of the United States Bureau of Education, "On the Training of Persons to Teach Agriculture in the Public Schools."

affairs of life; but I fear, however, that in the very prosperity of these redirected institutions there lies danger of undertaking kinds of work that partake overmuch of exploitation. Land and animals and orchards and machines and crops are no longer regarded as mere museums: they are laboratories and laboratory materials to be used for the same purpose and in the same pedagogical spirit as the geologist uses rocks or the chemist uses chemicals and chemical problems. We now have class-rooms into which cattle and sheep and other animals may be taken for study. These animals are laboratory material. If it is worth while to study live bacteria and live insects, it is equally worth while to study live cows. We have studied the fleas and other parasites that infest our domestic animals before we have studied the animals themselves, so successfully have we avoided the large and significant things.

In other words, the spirit of the modern agricultural college is to teach in terms of the actual daily life, making nature and the farm a real part of one's living and the foundation of one's philosophy of life. The lack of appreciation of this laboratory significance has prevented the proper growth of these agricultural institutions. They tell us that these colleges are now demanding enormous sums, but this is because we have never known how much money they have needed to make them effective. Never have they had money enough or freedom enough to work out their problem fundamentally. They are just beginning to develop. Agricultural education and experiment is the most expensive to maintain of all education because its laboratories are so large, so various, and so expensive in their up-keep. Institutions centering about city ideas receive no end of money and study. The open country is just coming to its own. Schools and colleges are worth only what they cost. With money and men working in state and national institutions, the rural problems can be solved.

It is strange that private benevolence has not discovered that the founding of schools of agriculture is one of the very best means of serving mankind. It is undoubtedly fortunate, however, that the people themselves have endowed these colleges, for this ensures that the institutions become and remain democratic, teachable and close to common problems. These colleges will place on the farms a class of educated persons, as the colleges of mechanic arts have placed such persons in the shops and in business.

The colleges of agriculture are essential because they are leading the way to a really useful training for country life. Our agricultural problem is one of constant readjustment to conditions, and this readjustment can progress only through the diffusion of greater intelligence. Knowledge and education lie at the very foundation of the welfare of the open country. Information and knowledge, however, and even education, do not of themselves constitute reform or progress. We need legislation and broad redirection of social and economic forces; but education lies behind and at the bottom of all these movements and without it no lasting progress is possible.

Interest in education by means of agriculture is no longer local. It is now more in the public mind than any other phase of education. It is interesting to note the zest with which the public is discovering the truths that the good prophets in the agricultural colleges announced ten and twenty years ago. The leadership in rural affairs is rapidly passing to the interests that associate themselves with the agricultural colleges and experiment stations. In twenty-five years there will be a new political philosophy of the open country born out of these institutions.

A mere enumeration of the departments comprising a modern college of agriculture indicates that while the main or central business of such college is to teach the science and the practice of farming, it really stands for the human affairs of the whole open country, taking this field because it is indivisible and also because other institutions have passed it by. There are institutions called universities that have a lesser scope than these leading colleges of agriculture, and in which the business affairs are less than in a modern dairy department of one of these colleges. These institutions mean not one iota less than the redirecting of the practices and ideals of country life, and they are today making the greatest single contribution to constructive pedagogical policies and for the very good reason that they deal with the commonplace facts and necessities of life. There was a day when universities tolerated instruction in agriculture. The time will soon be, if it is not already here, when a university that is a university must include agriculture.

The extension work of the colleges.

The extension effort is the most significant recent development of these colleges. It is an attempt to put the college in the way of aiding every man to help himself on his own farm. In this effort they have gone farther than any other institutions and they are setting an example for all institutions. Demonstration work, reading-courses, surveys, and similar enterprises, which are outgrowths of the colleges, are a part of this great extension movement. The extension work is the necessary distributive phase that aims, even though unconsciously, to overcome the effects of too much centralization at a distance. In spite of its crudity and formlessness, it is perhaps the directest effort yet made toward inspiring local initiative, and probably the best single contribution to the new social order of which I have been speaking.

The extension work of the agricultural colleges can hardly be said to need redirecting, because it has yet scarcely found its direction. Its purpose should be nothing less than to reach every farm and every farmer within its state or territory. Its purpose is not academic: it is service. But this work will find its greatest effectiveness and exercise economy of effort by dealing more with small groups of men than with isolated men, even if it is necessary to organize the groups in the first place. These groups will be represented in larger organizations. We have the germ of these larger groups in the "experiment unions" and similar organizations that are now arising in the agricultural colleges: in time these will be the greatest of farmers' associations, for they represent the point of view of the trained man. We may soon look for a larger federation of these state units, and the movement will then be nationalized (pages 115, 116).

Farmers' institutes will be one important part of this extension service. These institutes are now doing a great work, but a greater awaits them. They must be parts of organized educational centers. They must be fertilized by new and continuing study. They must be in the hands of specially trained men differing from both the college professor type and the so-called practical farmer type. We have not yet consciously trained such men. These institutes will fail of their greatest usefulness unless they coöperate fully with local organizations. In fact, it should be a part of their work to establish local organizations wherever they go, to continue and perfect the work. A reading-club for the systematic study of books and journals and bulletins should be the result of every institute.

These brief remarks on the colleges of agriculture and the experiment stations associated with them, indicate that these are not mere class institutions that are serving only technical farming needs. Yet there is constantly recurring comment in the press on the fact that not all the graduates return to farms and not all the bulletins reach farmers. These institutions are maintained by all the people, and for all the people who are interested in the work for which they stand. Neither do all graduates of colleges of law become lawyers, nor of colleges of medicine become physicians; nor is it desirable that they should. We need an educated laity in law and medicine, and equally in agriculture, for the big questions are social and national. The solution of the questions will not come in a day; but it is coming.

6. APPEAL TO PERSONAL LEADERSHIP

All these discussions mean that there needs to be a redirection of the point of view of the man. We have laid great emphasis on the necessity of making the farm more remunerative. Given the present income, however, whether in city or country, and it is possible for a man to lead a wholly different type of life. Much money is not essential with any of us to cleanliness, personal pride, sweetness of temper, honesty, and a few other very desirable but unpurchasable things. Most of all, the countryman needs intellectual horizon. He needs something else to think of. He needs to have a real personal sympathy with the natural objects in his environment. He needs the nature-study outlook. Whether a man wants much or little extraneous entertainment, or whether the country satisfies his ideals, depends on his attitude of mind.

With the great growth of urban sentiment and affairs, we have overlooked the value and significance of plain country living. Other ends in life have come into prominence, and persons have been attracted by the high points and by objects and affairs remote from them.

Real leadership lies in taking hold of the first and commonest problems that present themselves and working them out. Every community has its problems. Some one can aid to solve these problems. The size of the problem does not matter, if only some one takes hold of it and shakes it out. I like to say to my students that they should attack the first problem that presents itself when they alight from the train on their return from college. It may be a problem of roads; of a poor school; of tuberculosis in the herds; of ugly signs along the highways, where no man has a moral right to advertise private business; of a disease of apple trees; of poor seed; of the drainage of a field; of an improved method of growing a crop; of the care of the forests. Any young man can concentrate the sentiment of the community on a problem of the community. One problem solved or alleviated, and another awaits. The next school district needs help, the next town, the next county, the next state. Every able countryman has much more power than he uses.

Throughout this writing, I intend the word man to include also the farm woman: the language lacks a good word to imply them both. Women may be leaders in the large social work; and their influence should be paramount in redirecting the country home.

The scale of effort in the open country is so uniform that it ought to be easy to rise above it. I do not see how it is possible for an educated young man to avoid developing leadership in the open country, if only he attacks a plain, homely problem, is not above it, and sticks to it.

It does not follow that all leadership must be reached for. It will come to a man. A student recently asked my advice about his buying a farm. Since a mere boy he had desired to possess a certain farm in his neighborhood. It is a good farm, paying its owner a comfortable profit. The young man has no money and is working for his education. But now the owner of the farm is about to retire, and he offers the farm to this young man at a fair price. The young man can borrow the money and mortgage the place. He calculates that he can have it all paid for by the time he is forty-five, and the farm will turn a good living in the meantime. I think that the young man will make good. He knows farming. I advised him to buy. Most men on salary or in other business than farming do not have their homes paid for and their business all established at forty-five years of age. If the young man has the farm debt-free and fully productive at that age, and if he loves his neighbor as he loves his farm, he will not need to patronize anybody: persons will come to him. He cannot escape being a marked man. He should exemplify Milton's noble line,

"They also serve who only stand and wait."

In this brief sketch, then, I have tried to show that the rural country needs a new direction of effort, a new outlook, and a new inspiration. Every rural institution should have direct relation to the land on which it stands. Education should take hold of every factor that means much to the people. Some man some day will see the opportunity and will seize it. The result of his work will be

Leadership Waits for One   דרך

simply a new way of thinking; but it will eventuate into a new political and social economy. When his statue is finally cast in bronze, he will not be placed on a prancing steed nor surrounded by any symbols of carnage or of war. He will be a plain man in citizen's clothes and he will stand on the ground; but his face will be toward the daylight.

LC

www.ingramcontent.com/pod-product-compliance
Lightning Source LLC
Chambersburg PA
CBHW082327220526
45470CB00008B/2427